NUMERACY ESSENTIALS

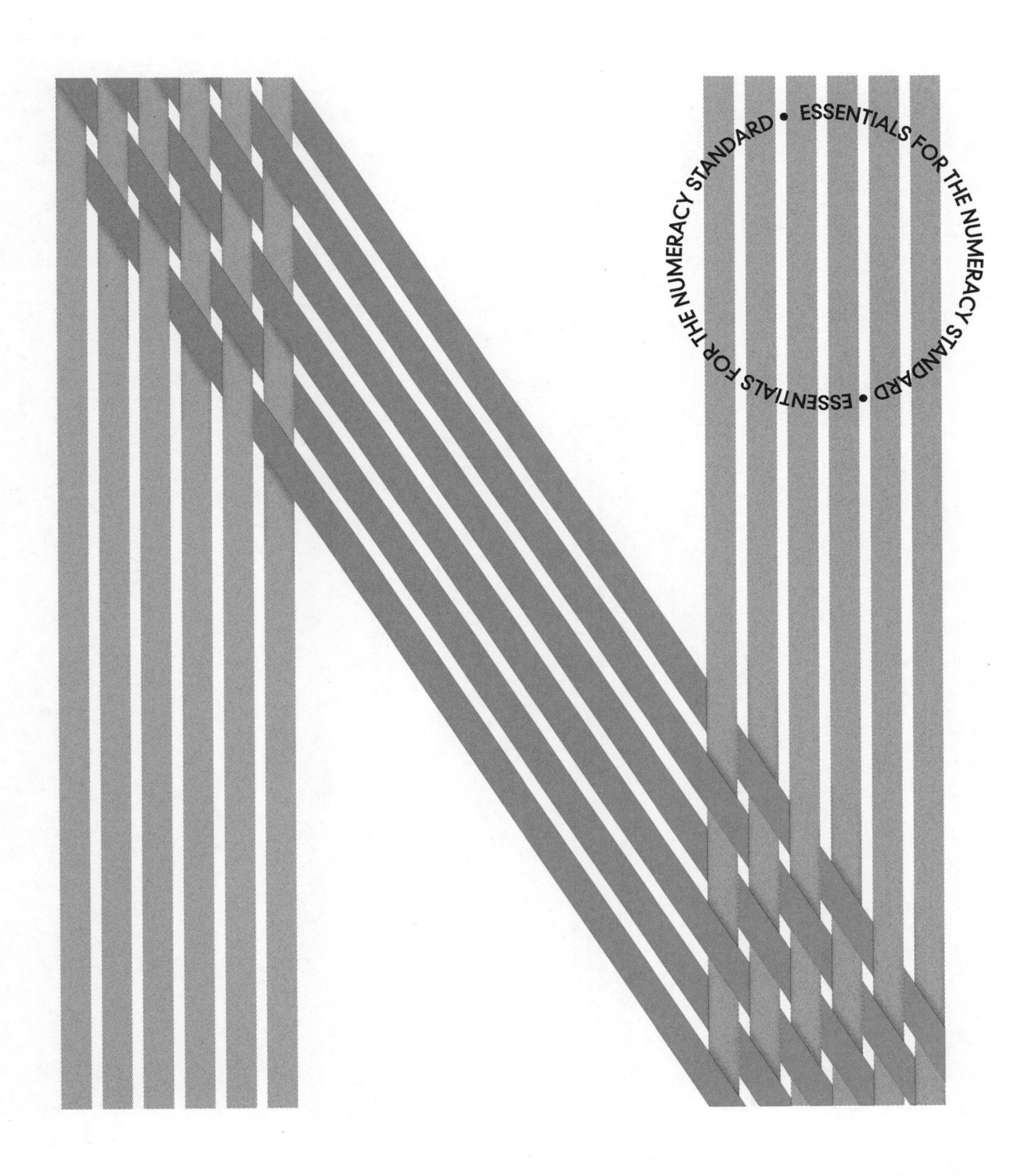

Victoria and Charlotte Walker

Walker Maths: Numeracy Essentials
1st Edition
Charlotte Walker
Victoria Walker

Designer: Cheryl Smith, Macarn Design
Production controller: Katie McCappin

Acknowledgements
The authors wish to thank past and present colleagues who have generously shared their expertise and ideas.

For product information and technology assistance,
in Australia call **1300 790 853**;
in New Zealand call **0800 449 725**

For permission to use material from this text or product, please email **aust.permissions@cengage.com**

National Library of New Zealand Cataloguing-in-Publication Data
A catalogue record for this book is available from the National Library of New Zealand.

978 0 17 047771 0

Cengage Learning Australia
Level 5, 80 Dorcas Street
Southbank VIC 3006 Australia

Cengage Learning New Zealand
For learning solutions, visit **cengage.co.nz**

Printed in China by 1010 Printing International Limited.
4 5 6 7 27 26 25 24

CONTENTS

Operations on numbers

Place value

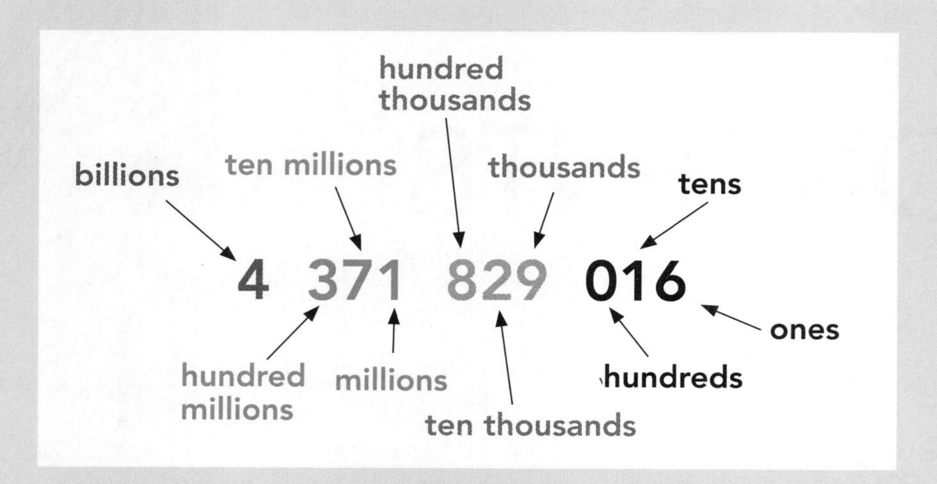

- When converting numbers written using numerals (or digits) to words, split the number where the gaps occur.

Examples:

1 Write 15 027 403 021 using words.

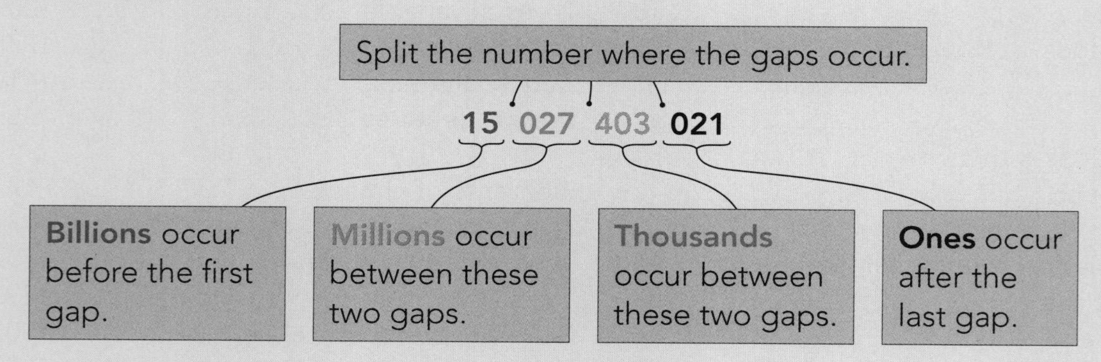

Fifteen billion, **twenty-seven million**, **four hundred and three thousand**, **and twenty-one**.

2 Write eighteen billion, three hundred and sixty-five million, one hundred and ninety-seven thousand, four hundred and one using digits.

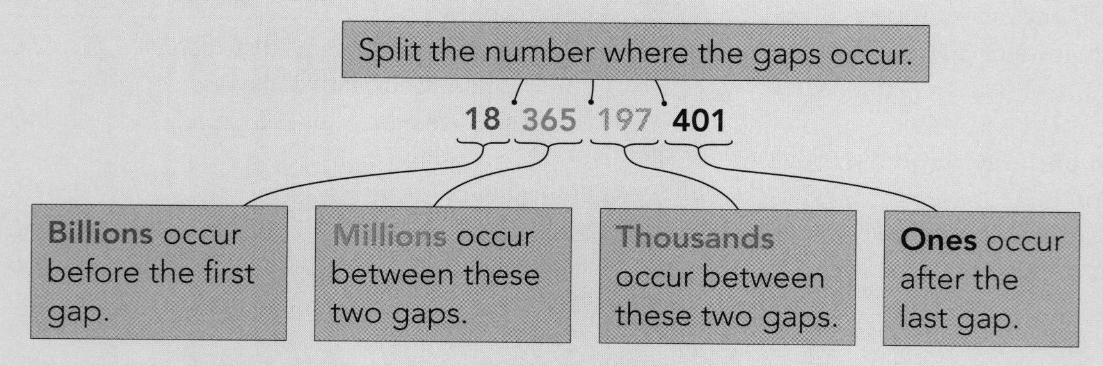

Eighteen billion, **three hundred and sixty-five million**, **one hundred and ninety-seven thousand**, **four hundred and one**.

ISBN: 9780170477710

1 Join the dots to connect each number written using words with the same number written using numerals.

One hundred	•	• 1 000 000
One billion	•	• 1 000
One million	•	• 100
Ten million	•	• 100 000 000
One hundred million	•	• 10 000
One thousand	•	• 10 000 000
One hundred thousand	•	• 1 000 000 000
Ten thousand	•	• 100 000

Write down the value of the green number using numerals and words.

		Numerals	Words
2	23 **4**91	400	
3	1 2**7**2 408		Seventy thousand
4	**7** 419 364 201		
5	44**2** 196 939		
6	**6**75 521		
7	**9**13 841 725		
8	43**1** 612		
9	1**8**2 341 092		

N

10 Match these values to the words below.

2 005 840	20 584	2 000 005 840
2 500 800 040	2 500 804	205 840
250 084	2 584	2 050 840

Two thousand, five hundred and eighty-four ________________	Two million, five hundred thousand, eight hundred and four ________________	Two billion, five thousand, eight hundred and forty ________________
Two million, fifty thousand, eight hundred and forty ________________	Twenty thousand, five hundred and eighty-four ________________	Two hundred and five thousand, eight hundred and forty ________________
Two billion, five hundred million, eight hundred thousand and forty ________________	Two million, five thousand, eight hundred and forty ________________	Two hundred and fifty thousand and eighty-four ________________

Write these numbers using words.

11 82 137 ________________________________

12 8 010 300 ________________________________

13 204 009 ________________________________

14 2 001 032 000 ________________________________

15 901 014 561 ________________________________

Write these numbers using numerals.

16 Sixty-four thousand, nine hundred ________________

17 Seven hundred and ninety-nine thousand ________________

18 One million, six hundred thousand and thirty ________________

19 Seven hundred thousand, two hundred and nine ________________

20 Eleven million, four thousand, nine hundred and eighty ________________

21 One billion, three hundred and fifty million ________________

ISBN: 9780170477710

Rounding to whole numbers

- Often we need to round numbers to sensible and/or meaningful values.
- Never round until **after** you have completed your calculations.

Locate the first digit to the **right** of last required digit. Use this digit **only**, ask yourself:

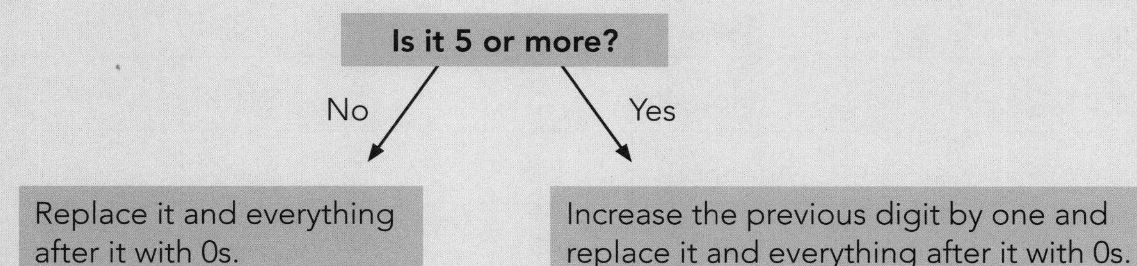

Examples:

1 Round 236 to the nearest ten.

Is 236 closer to 230 or 240?

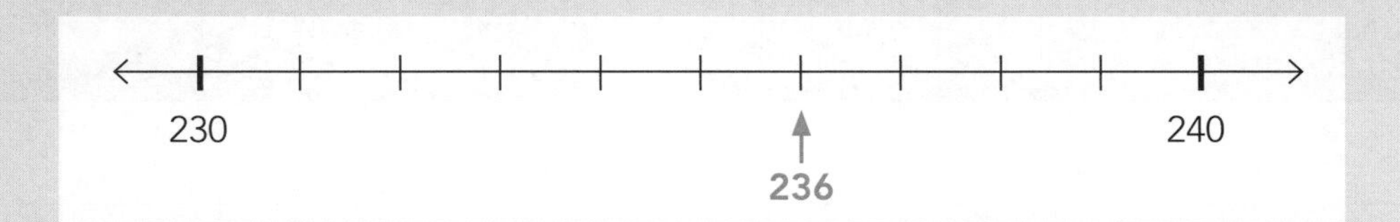

The digit to the right of the last required digit in 236 is the **6**.
So 236 rounded to the nearest ten = 240.

2 Round 1 649 to the nearest hundred.

Is 1 649 closer to 1 600 or 1 700?

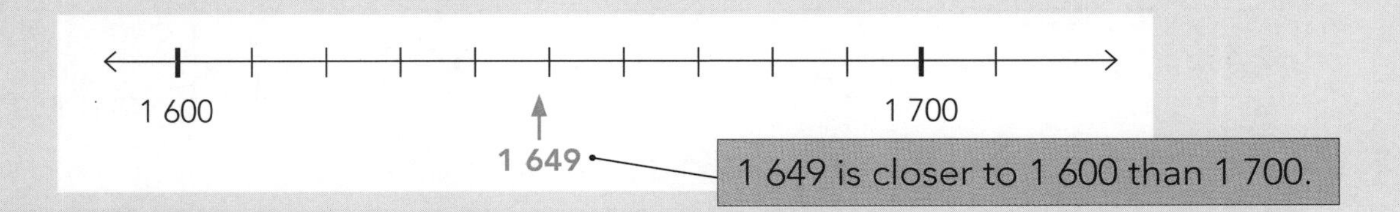

The digit to the right of the last required digit in 1 649 is the **4**.
So 1 649 rounded to the nearest hundred = 1 600.

3 Round 31 500 to the nearest thousand.

Is 31 500 closer to 31 000 or 32 000?

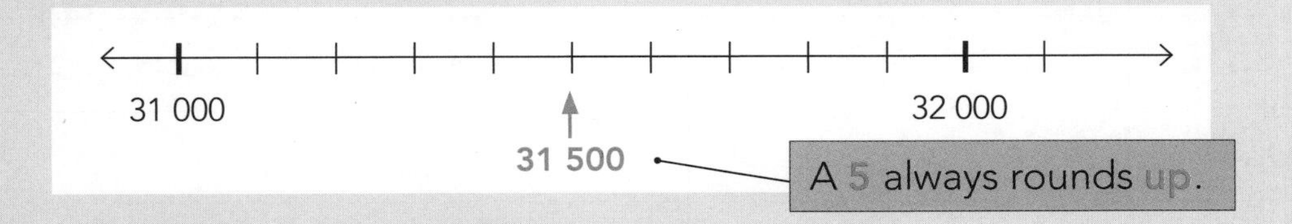

The digit to the right of the last required digit in 31 500 is the **5**.
So 31 500 rounded to the nearest thousand = 32 000

More examples:

	Rounded to the nearest:	Digit to the right of the last required digit	Answer
29	ten	29	30
3 457	hundred	3 457	3 500
841 273	thousand	841 273	841 000
1 869 204	ten thousand	1 869 204	1 870 000
4 790 351	million	4 790 351	5 000 000

1 Complete this table.

	Rounded to the nearest:	Highlight the digit to the right of the last required digit	Answer
518	ten	518	520
72 538	thousand	72 538	
4 197	hundred	4 197	
634 116	ten thousand	634 116	
2 457 101	million	2 457 101	
67 452	hundred	67 452	

2 Using the numbers in the left column, complete this table.

	Round to the nearest ten	Round to the nearest hundred	Round to the nearest thousand
7 813	7 810	7 800	8 000
23 465			
817 337			
1 949			
96 726			
3 455 752			
531 425			

 ISBN: 9780170477710

Round these numbers to the nearest ten.

3 37 ______________ **4** 831 ______________

5 16 456 ______________ **6** 5 149 ______________

Round these numbers to the nearest hundred.

7 424 ______________ **8** 13 641 ______________

9 5 371 093 ______________ **10** 712 245 ______________

Round these numbers to the nearest thousand.

11 1 706 ______________ **12** 185 490 ______________

13 854 ______________ **14** 27 439 ______________

Round these numbers to the nearest ten thousand.

15 52 471 ______________ **16** 64 328 674 ______________

17 4 116 431 ______________ **18** 111 531 ______________

Round these numbers to the nearest million.

19 13 256 989 ______________ **20** 1 562 473 ______________

21 56 389 041 ______________ **22** 9 790 004 ______________

23 Complete the table.

Original number	Rounded number	This was rounded to the nearest:
947	950	
8 276	8 000	
12 846	12 850	
367 482	267 500	
452 791	450 000	
24 613 094	24 600 000	
490 578 935	500 000 000	
1 642 542 895	2 000 000 000	

N

Integers

Adding and subtracting

- This can be done on a **number line**.
- – means move **left**.
- + means move **right**.
- Remember that – – = +

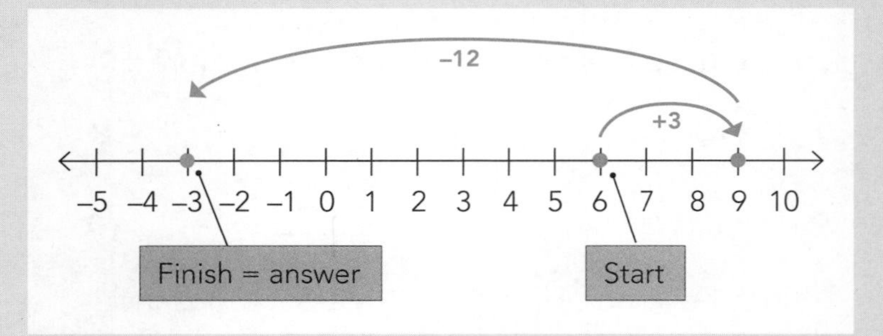

Example: 6 – (–3) – 12 = 6 + 3 – 12
= –3

– – ⇒ +

Add arrows and dots to these number lines in order to complete the calculations.

1 3 – 7 =

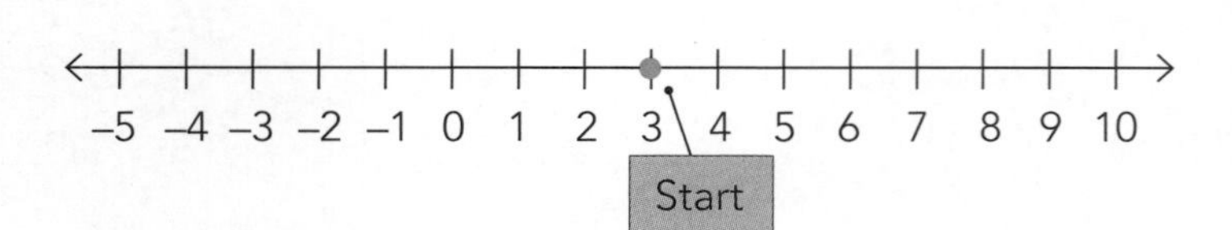

2 –2 – 3 + 11 =

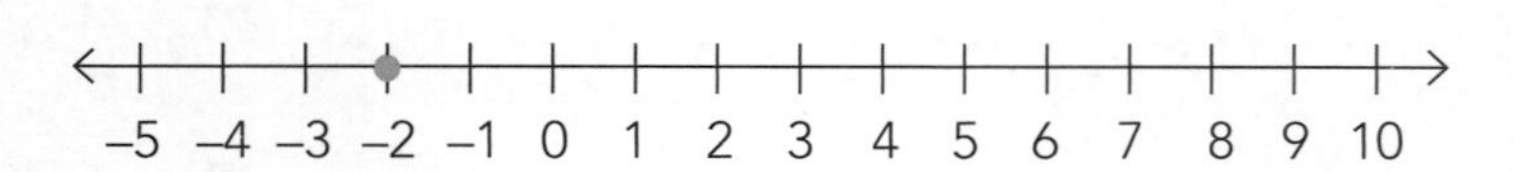

3 –1 – 5 – (–2) =

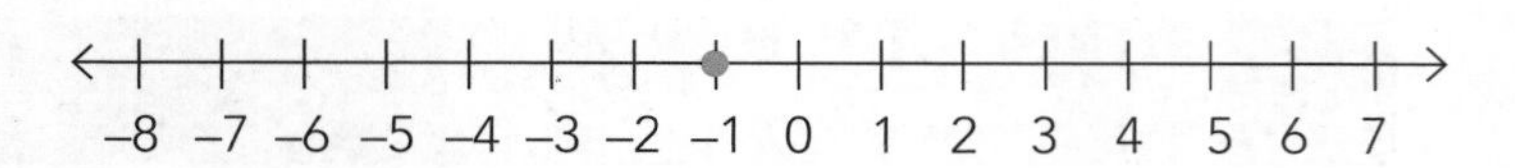

4 –40 + 55 – 70 =

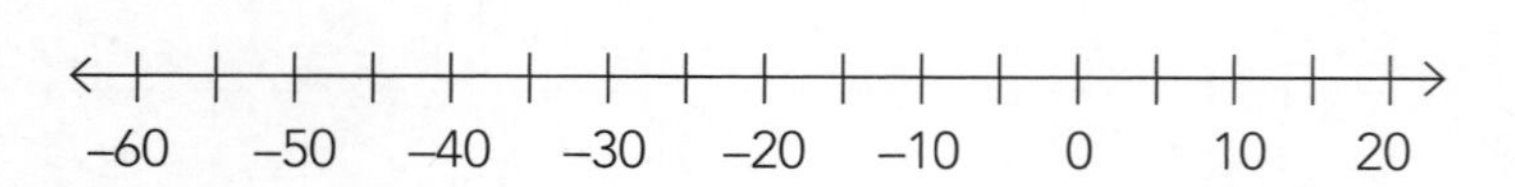

Use your calculator to complete these calculations.

5 –54 + 18 = ____________

6 3 – 12 = ____________

7 –5 + 8 – 9 = ____________

8 14 – (–5) – 21 = ____________

9 –140 + 35 + 15 = ____________

10 450 – 1009 + 345 = ____________

ISBN: 9780170477710

Place these values in ascending order (smallest to largest). Cross out each value as you go.

11 19 –9 0 –1 ~~9~~ –19 –10 10

Smallest | | | | | | | Largest

					9		

12 –101 0 –110 101 –10 –1 110 –111

Smallest | | | | | | | Largest

13 Write the integer values for each point along the number line. Choose the most appropriate values from the options below. You will not need all of them.

–8	–19	–5	–17	–18	–3
–1	–14	1	–4	–11	5
–10	–6	–15	–7	3	–22

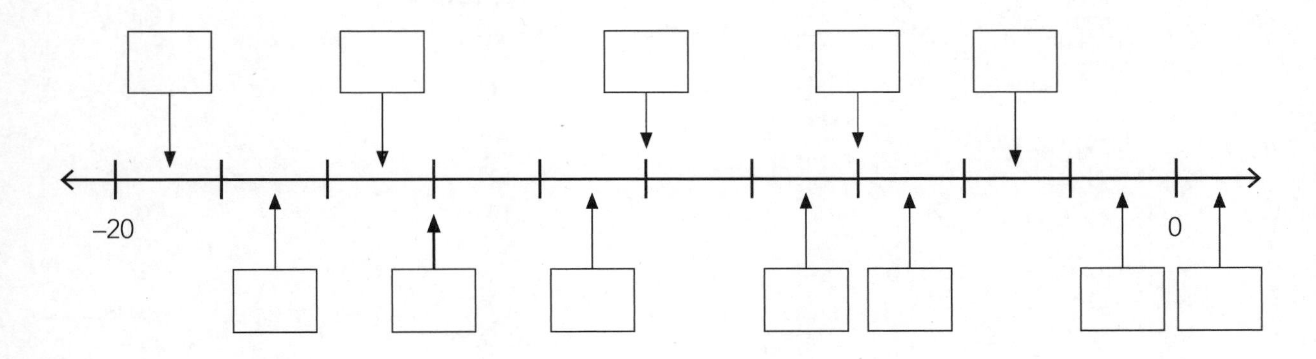

14 The temperature was 5°C. By dusk it had dropped by 6°C and by dawn it dropped another 5°C. What was the temperature at dawn?

_______°C

15 Annie owed her mum $17 and her brother $25. She was paid $79, then while walking home found a $5 note in the park. She then repaid her debts to her mum and brother. How much did she have left?

$_________

16 Eru's freezer was at –18°C. There was a power cut during which the freezer temperature rose by 5°C. The power came back on and over the next hour the freezer temperature dropped by 2°C. What was the freezer temperature at the end of this hour?

_______°C

ISBN: 9780170477710

N

Powers

- Powers are used to indicate how many times a number (the base) is multiplied by itself.

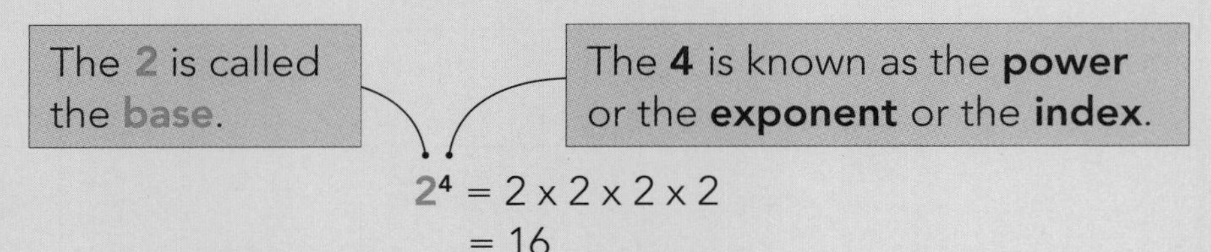

Example: $2^4 = 2 \times 2 \times 2 \times 2$
$= 16$

- When a power is 2, we say the number is **squared**, e.g. 4^2 means four **squared** (= 16).
- When a power is 3, we say the number is **cubed**, e.g. 2^3 means two **cubed** (= 8).

The **power** indicates how many of these numbers you need to multiply together.

Examples: $6^2 = 6 \times 6$
$= 36$

$4^5 = 4 \times 4 \times 4 \times 4 \times 4$
$= 1024$

Finding powers on your calculator

- Powers of numbers can get really big, so knowing how to find them on your calculator is **very** useful.

For squares: use a button that looks like this: x^2 e.g. show that $15^2 = 225$.

For all other powers: use the button that looks like this: $x^{\blacksquare}$ e.g. show that $2^6 = 64$.

Write the following as powers.

1 $5 \times 5 \times 5 \times 5$ = 5^4

2 3×3 = ____________

3 $8 \times 8 \times 8$ = ____________

4 $7 \times 7 \times 7 \times 7$ = ____________

Write out the meaning of these expressions and then use your calculator to find their values.

5 2^4 = $2 \times 2 \times 2 \times 2$ = ________

6 3^5 = ________________________ = ________

7 6^3 = ________________________ = ________

8 9^4 = ________________________ = ________

Use your calculator to find the values of these expressions.

9 $2^3 \times 2^2$ = ________________________

10 $3^3 \times 10^2$ = ________________________

11 $5^4 \times 2^3$ = ________________________

12 $4^2 \times 10^5$ = ________________________

ISBN: 9780170477710

Words to calculations

1 At the end of the year, Noa, and Amy bought their maths teacher a box of chocolates which cost $12 and a card which cost $3. If they shared the cost, how much did they each pay?

____

______________________________ $______

2 Two brothers went to a concert. Their tickets cost $15 each and they shared a $4 box of popcorn. How much did it cost each brother?

____

______________________________ $______

3 Three friends bought a bundle containing 24 sacks which cost them $18. They filled the sacks with pine cones and sold the filled sacks for $5 each. They shared their profits equally. How much did each friend earn?

____

______________________________ $______

4 Tane was given $16 for mowing the lawn. He paid his sister the $3.60 he owed her. He then spent half. How much money did Tane have left?

____

______________________________ $______

5 Manu and Henry stacked firewood for Manu's kuia. She paid them $25 altogether. On the way home they bought a $4 lolly mix each. They shared the remaining money. How much did each get?

____

______________________________ $______

6 Sam and Anna sold 28 plants for $3 each at their roadside stall. The pots had cost them $8.40. They shared the profit equally. How much did each get?

____

______________________________ $______

7 A family of two adults and two children went to the zoo. It cost the adults $28 each and children $14 each. They also bought two ice creams at $3.50 each. How much did it cost in total?

____

______________________________ $______

8 Huia worked for five hours at $18 per hour. On her way home she bought a $4.80 milkshake. Then she gave her mum half of what remained. How much did she give her mum?

____

______________________________ $______

N

Decimals

- Decimals are the values that come after the decimal point.
- Calcuations with money use decimals.

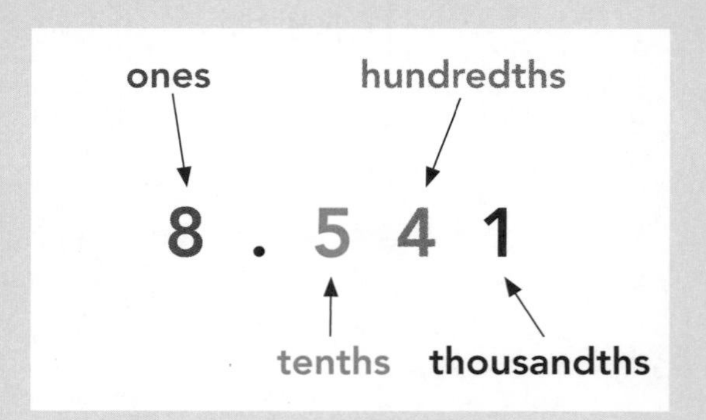

Comparing decimals

- To determine which decimal is largest, you need to consider the place values of each digit, starting from the left.

Examples: Identify the larger number.

1 34.6 34.7

Step 1: Arrange the numbers with the decimal points lined up vertically:

34.6
34.7

The decimal points must be in line.

Step 2: Start at the left, and look for the first pair of digits that are different:

The first two digits are the same.

34|.6
34|.7

Step 3: Decide which is greater: 6 or 7? 7 is larger so 34.7 is larger than 34.6.

2 19.23 19.21

Step 1: Arrange the numbers with the decimal points lined up vertically:

19.23
19.21

Step 2: Start at the left, and look for the first pair of digits that are different:

The first three digits are the same.

19.2|3
19.2|1

Step 3: Decide which is greater: 3 or 1? 3 is larger so 19.23 is larger than 19.21.

Highlight the larger number.

1	19.7	19.6	**2**	1651.3	1651.2
3	10.61	10.62	**4**	98.98	98.97
5	0.90	0.09	**6**	0.127	0.137

ISBN: 9780170477710

Place these decimals in ascending order (from smallest to largest).

7 3.24, 2.34, 3.42, 2.43 ________ ________ ________ ________

8 0.60, 0.61, 0.06, 0.16 ________ ________ ________ ________

9 9.10, 9.01, 9.11, 9.00 ________ ________ ________ ________

10 0.101, 0.110, 0.011, 0.010 ________ ________ ________ ________

11 Write decimal values for each point along the number line. Choose the most appropriate values from the options below. You will not need all of them.

0.3	1.2	0.6	0.9
2.1	0.2	0.4	0.7
0.1	1.1	0.5	0.8

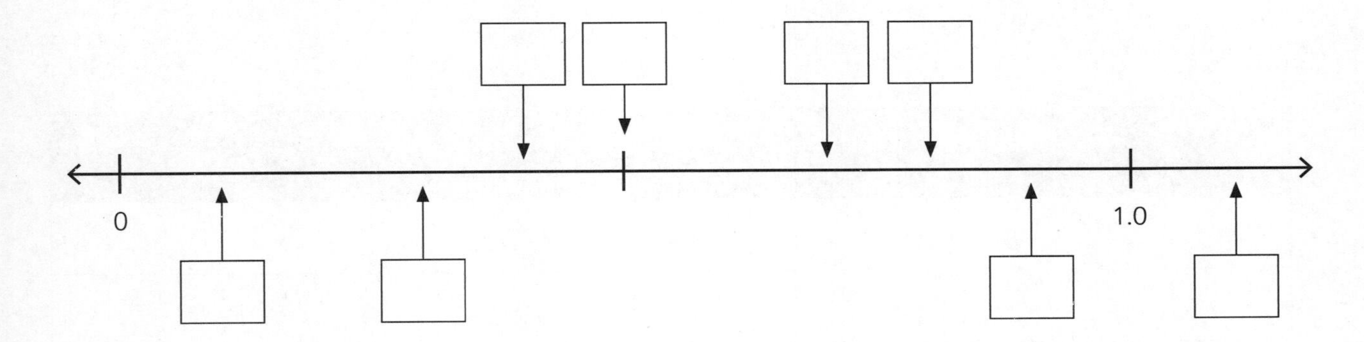

12 Write decimals values for each point along the number line. Choose the most appropriate values from the options below. You will not need all of them.

0.95	0.26	0.62	0.35	2.01
0.83	0.81	1.07	0.71	1.17
1.70	0.99	0.53	0.75	0.92

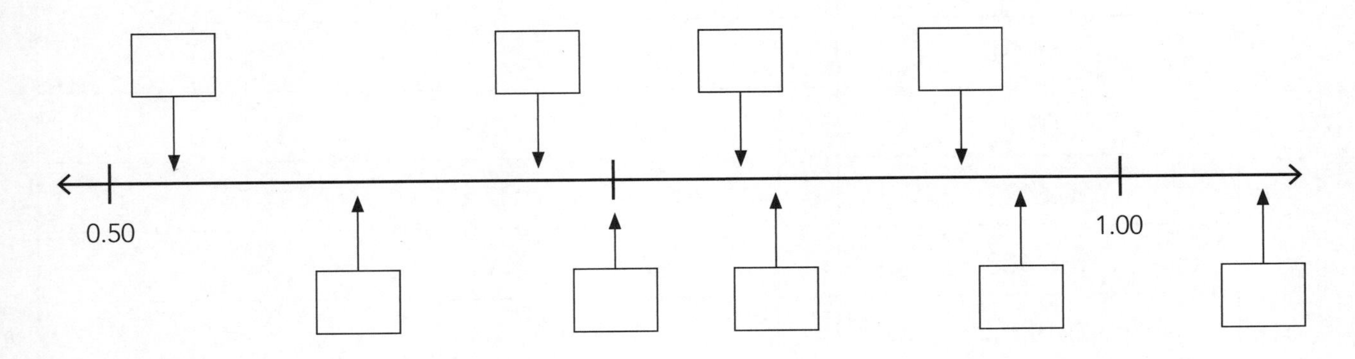

ISBN: 9780170477710

N

Rounding decimals

- Often we need to round numbers to sensible and/or meaningful values.
- Never round until **after** you have completed your calculations.
- The number of decimal places is the number of digits after the decimal point.
- When rounding, the decimal point **does not** move.

Examples:

Also known as rounded to the nearest tenths.

Number	6	6.2	6.17	6.170
Number of decimal places	0	1	2	3

This is the same as rounding to the nearest whole number.

Notice that a **0** at the **end** counts as a decimal place.

- When asked to round to 1 dp (one decimal place), there should be **exactly** one digit after the decimal point.
- The process is the same as for rounding to whole numbers.

Examples:

	Rounded to the nearest:	Digit to the right of the last required digit	Answer
1.372	1 dp	1.372	1.4
32.579	2 dp	32.579	32.58
512.7404	3 dp	512.7404	512.740

13 Complete the table.

Original number	Rounded number	Rounded to how many dp?
4.814	4.81	
95.571	95.6	
251.19	251	
1.0935	1.094	
0.617	0.6	
168.408	168.41	
0.1996	0.20	

ISBN: 9780170477710

14 Complete the table.

	Rounded to the nearest:	Highlight the digit to the right of the last required digit	Answer
29.372	1 dp	29.372	
1.0375	2 dp	1.0375	
132.6724	3 dp	132.6724	
10.642	0 dp	10.642	
9566.353	2 dp	9566.353	
0.3914	1 dp	0.3914	
4.5271	3 dp	4.5271	
0.0089	2 dp	0.0089	

15 Complete the table.

	0 dp	1 dp	2 dp
8.9421			8.94
16.738			
9.3072		9.3	
1.0997	1		
0.7254		0.7	
284.042			

Do these calculations and then round the answer to the required number of decimal places.

16 0.56 x 0.23 = ____________ (2 dp)

17 1.2 x 0.68 = ____________ (1 dp)

18 10.05 x 2.5 = ____________ (2 dp)

19 0.88 x 1.5 x 0.49 = ____________ (3 dp)

20 (3.61 + 1.8) x 0.25 = ____________ (3 dp)

21 18.4 ÷ 30 = ____________ (3 dp)

22 $\frac{4.165}{17}$ = ____________ (2 dp)

23 $\frac{244.5 + 6.35}{86.5}$ = ____________ (0 dp)

Fractions

- Fractions are another way of writing numbers or parts of numbers
- Remember that 1 is a whole: it could also be written $\frac{2}{2}$ or $\frac{5}{5}$, for example.

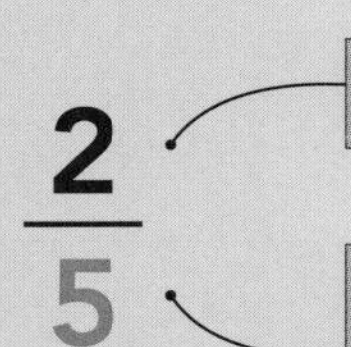

The **top** number is called the **numerator**.

The **bottom** number is called the **denominator**.
Hint: d is also the first letter of downstairs.

Examples:

$\frac{1}{5}$ 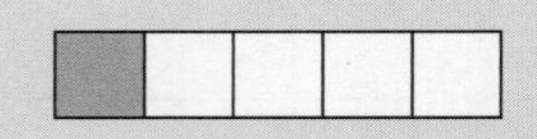 $\frac{2}{5}$ 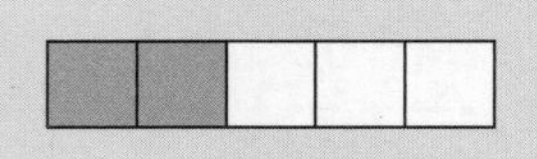

When the **numerator increases**, the size of the shaded section **increases**.
So $\frac{2}{5}$ is larger than $\frac{1}{5}$.

$\frac{1}{3}$ 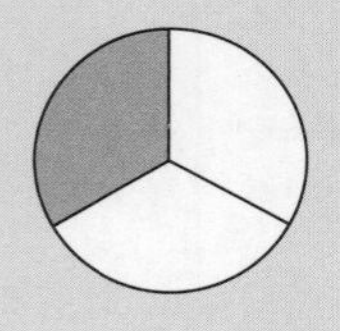 $\frac{1}{4}$ 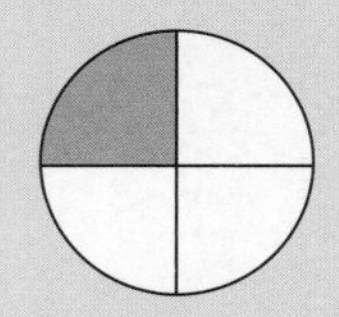

When the **denominator increases**, the size of the shaded section **decreases**.
So $\frac{1}{3}$ is larger than $\frac{1}{4}$.

Highlight the larger number.

1 $\frac{1}{6}$ $\quad$ $\frac{1}{8}$

2 $\frac{2}{7}$ $\quad$ $\frac{3}{7}$

3 $\frac{1}{12}$ $\quad$ $\frac{1}{9}$

4 $\frac{5}{6}$ $\quad$ $\frac{4}{6}$

5 $1\frac{1}{3}$ $\quad$ $1\frac{1}{4}$

6 $\frac{5}{5}$ $\quad$ $\frac{4}{5}$

- You can also compare fractions by converting them to decimals and then comparing the decimals.

Example: $\frac{3}{20} = 0.15$ and $\frac{1}{8} = 0.125$, so $\frac{3}{20}$ is larger than $\frac{1}{8}$.

To use your calculator: 3  20 = 

Convert the following fractions to decimals using your calculator, and decide which is larger.

7 $\frac{3}{5}$ = ________ $\quad$ _____ is larger
$\frac{5}{8}$ = ________ $\quad$ than _____

8 $\frac{1}{3}$ = ________ $\quad$ _____ is larger
$\frac{7}{20}$ = ________ $\quad$ than _____

9 $1\frac{1}{5}$ = ________ $\quad$ _____ is larger
$1\frac{3}{25}$ = ________ $\quad$ than _____

10 $\frac{3}{4}$ = ________ $\quad$ _____ is larger
$4\frac{29}{40}$ = ________ $\quad$ than _____

ISBN: 9780170477710

Fraction calculations

- You need to understand how to use your calculator to find answers to problems which use fractions.
- Throughout this book there are instructions in the use of calculators. Your calculator may be different, so you might have to experiment to find what works.

Some useful calculator buttons might look like  and .

Use your calculator to calculate these. Write your answers as fractions.

11 $\frac{7}{8} - \frac{1}{3} =$ __________

12 $1\frac{1}{2} + \frac{2}{5} =$ __________

13 $2\frac{1}{3} - \frac{1}{7} =$ __________

14 $7\frac{4}{5} - 3\frac{1}{4} =$ __________

15 $2\frac{3}{4} \times \frac{1}{2} =$ __________

16 $4\frac{1}{2} \times 1\frac{1}{4} =$ __________

17 $5\frac{3}{5} \div 7 =$ __________

18 $10\frac{3}{4} \div \frac{1}{2} =$ __________

Use your calculator to calculate these. Write your answers as decimals.

19 $\frac{3}{8} + \frac{1}{5} =$ __________

20 $7\frac{9}{40} - \frac{3}{10} =$ __________

21 $11 - 1\frac{7}{8} =$ __________

22 $7\frac{4}{5} + \frac{3}{4} - 1\frac{1}{2} =$ __________

23 $15\frac{3}{4} \div 9 =$ __________

24 $2\frac{1}{2} \times 1\frac{1}{2} =$ __________

25 Hemi ate $\frac{1}{4}$ of the pizza, Tom ate two fifths, and Jake ate the rest. What fraction of the pizza did Jake eat? __________

26 Rawiri's scone recipe needs $1\frac{1}{4}$ cups of milk. He has $\frac{2}{3}$ of a cup. How much more milk does he need?
Give your answer as a fraction of a cup. __________

ISBN: 9780170477710

N

Finding a fraction of an amount

- Remember that '**of**' means you must **multiply**.

Example: $\frac{1}{4}$ of 36 = $\frac{1}{4} \times 36$ or $\frac{1}{4}$ of 36 = $1 \div 4 \times 36$

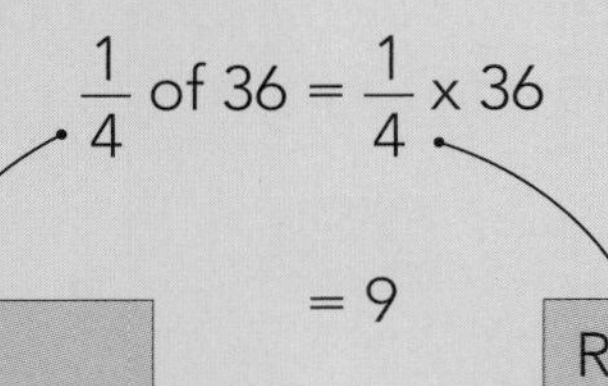

= 9 = 9

Use your fraction button.

Replace the word '**of**' with a **x** sign.

Answer the following questions.

27 $\frac{1}{3}$ of 48 = ____________

28 $\frac{2}{5}$ of 90 = ____________

29 $\frac{5}{6}$ of 75 = ____________

30 $\frac{1}{6}$ of 33 = ____________

31 $\frac{1}{7}$ of 98 = ____________

32 $\frac{2}{9}$ of 135 = ____________

33 $\frac{14}{15}$ of 165 = ____________

34 $1\frac{1}{4}$ times 64 = ____________

35 $4\frac{2}{3}$ times 15 = ____________

36 $10\frac{1}{4}$ times 60 = ____________

37 Five friends ate dinner at a restaurant. The bill totalled \$142.50. They split the bill equally. How much did each friend pay? \$__________

38 Kara bought a bag of jellybeans. It contained 128 jellybeans. An eighth of these were her favourite colour (black). How many black jellybeans were there? __________

39 Four sevenths of the students walk to school. There are 532 students in total. How many students walk to school? __________

40 The number of people in New Zealand was about one fifth of the number of sheep in 2022. If there were about 25.97 million sheep in 2022, about how many people were there? __________

ISBN: 9780170477710

Percentages

- Percentages are a way of expressing a number **out of 100**.

Example:

32 out of 100 squares are green.

$\frac{32}{100}$ is the same as 32%.

$\frac{68}{100}$ of 100 squares are white.

This means 68% are white.

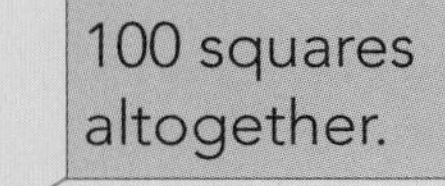

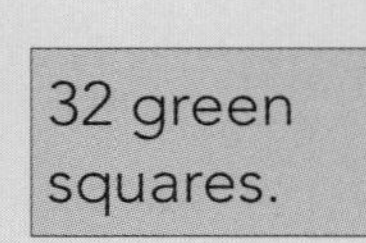

100 – 32 = 68
white squares.

Sometimes there are not 100 squares.

Example:

41 out of 50 squares are green.

$\frac{41}{50}$ needs to be multiplied by 100 to make a percentage.

$\frac{41}{50}$ **x 100** = 82%, so 82% of the squares are green.

$\frac{9}{50}$ of the squares are white.

$\frac{9}{50}$ **x 100** = 18%, so 18% of the squares are white.

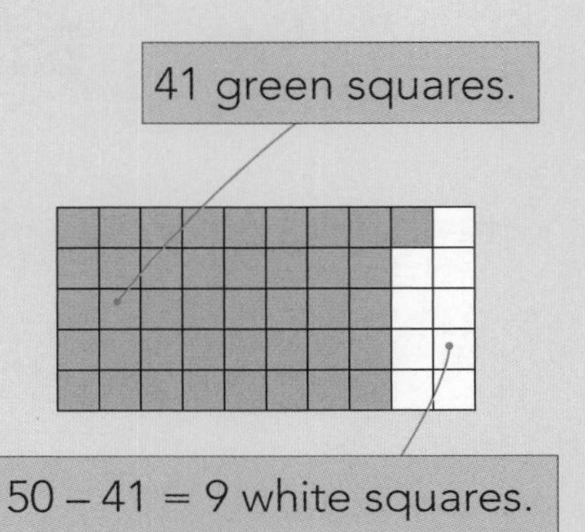

State the percentages shown in these diagrams.

1

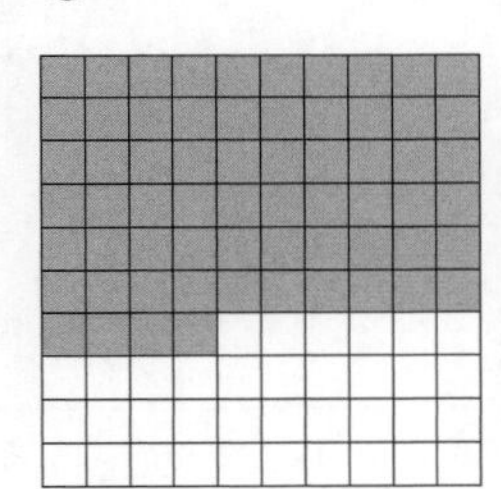

Percentage green ______________

Percentage white ______________

2

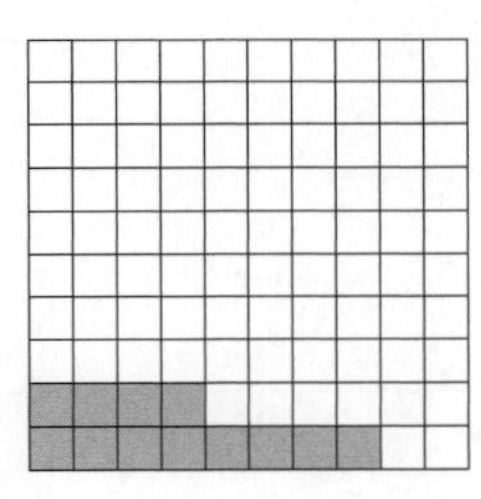

Percentage green ______________

Percentage white ______________

3

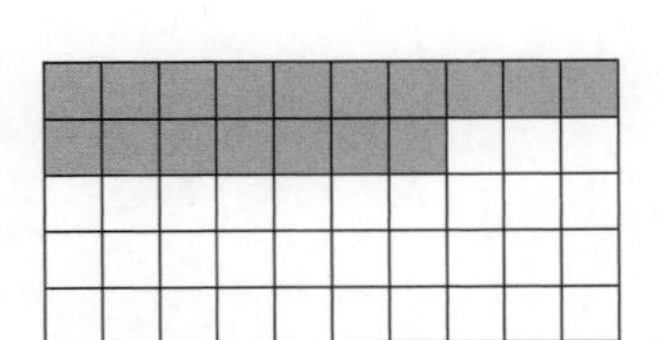

Percentage green $\frac{\quad}{50}$ x 100 = ______

Percentage white $\frac{\quad}{50}$ x 100 = ______

4

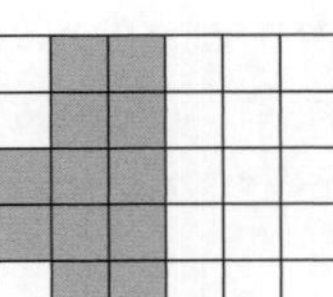

Percentage green — x 100 = ______

Percentage white — x 100 = ______

ISBN: 9780170477710

N

Calculating percentages

- Remember, 'per cent' means out of 100, so turning a fraction into a percentage means **multiplying by 100**.

Example: There are 30 students in Mia's class; 12 of them are boys. What percentage of the class are boys?

$$\text{Percentage} = \frac{12}{30} \times 100 = 40\%$$

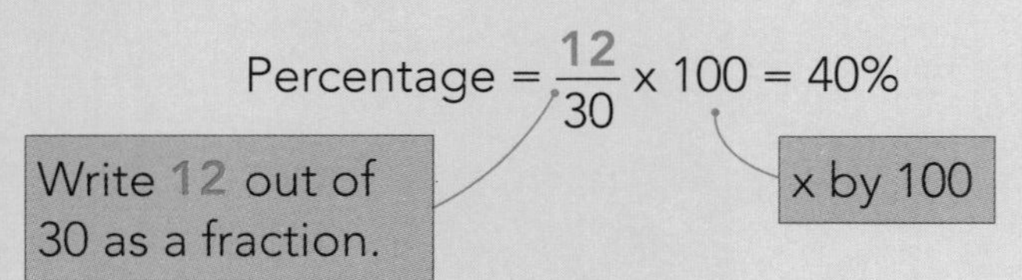

Otherwise use your calculator.

12 ÷ 30 

=

Pressing the shift button means your calculator will do the operation on the body, not the one on the key.

Write the following amounts as percentages.

5 17 out of 40 = ______________________
= ______________________

6 48 out of 120 = ______________________
= ______________________

7 120 out of 150 = ______________________
= ______________________

8 1998 out of 3000 = ______________________
= ______________________

9 $\frac{7}{8}$ = ______________________
= ______________________

10 $\frac{23}{25}$ = ______________________
= ______________________

11 $\frac{105}{112}$ = ______________________
= ______________________

12 $\frac{36}{225}$ = ______________________
= ______________________

13 Ethan got 38 out of 40 questions correct in a test. What percentage was this? ____________%

14 Thirteen out of the world's eighteen penguin species have been seen in New Zealand. What percentage of the world's species does this represent? Round your answer to 1 dp. ____________%

15 The earth is made up of 35% iron, 30% oxygen, 15% silicon and 13% magnesium. What percentage is made up of other elements? ____________%

ISBN: 9780170477710

Converting between decimals, fractions and percentages

- It's useful to be able to convert between decimals, fractions and percentages.

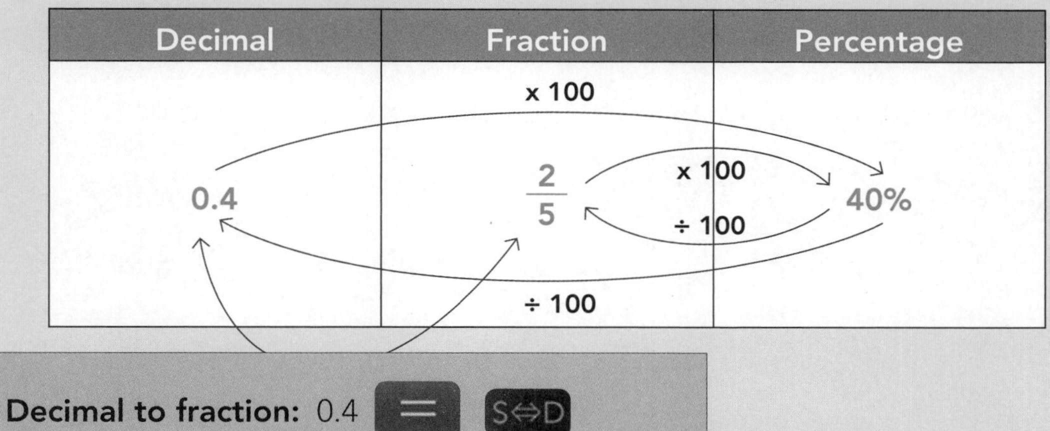

Decimal to fraction: 0.4 [=] [S⇔D]

Fraction to decimal: 2 [fraction] 5 [=] [S⇔D]

16 Fill in the gaps.

Decimal	Fraction	Percentage
	$\frac{1}{2}$	
0.6		
		25%
	$\frac{1}{5}$	
0.4		
		80%
0.75		
	$\frac{1}{3}$	
		$66.\dot{6}\%$

ISBN: 9780170477710

N

Finding percentages of amounts

- There are several ways of doing this.

Example: Find 35% of 80.

Either: convert the percentage to a decimal: 35% **of** 80 = 0.35 **x 80**

= 28

Remember, '**of**' means **x**.

Or: convert the percentage to a fraction: 35% **of** 80 = $\frac{35}{100}$ **x 80**

= 28

Otherwise use your calculator.

80  25 

Calculate the following.

17 10% of 60 = ____________________

= ____________________

18 25% of 92 = ____________________

= ____________________

19 20% of 25 = ____________________

= ____________________

20 60% of 140 = ____________________

= ____________________

21 8% of 234 = ____________________

= ____________________

22 2% of 5 600 = ____________________

= ____________________

23 About 10% of people are left handed. If a school has 739 students on its roll, about how many students would you expect to be left handed?

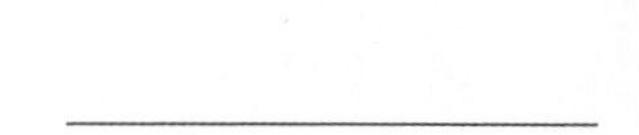

24 Water covers 70% of the earth's surface. The total surface area of the earth is 510.1 million km^2. How much of the earth is covered by land?

 km^2

25 In 2022 there were about 3.3 million light vehicles. About 1.15% of these were electric. About how many electric vehicles were there in 2022?

ISBN: 9780170477710

Increasing by a percentage

- There are several ways of doing this.

Example: Increase 86 by 35%.

This means that we need 100% **plus** 35%.

Increased amount = 86 + 35% of 86
= 86 + 30.1
= 116.1

or

Increased amount = 86 x 135%
= 86 x 1.35
= 116.1

Otherwise use your calculator.

86  135   

Calculate the following.

26 Increase 60 by 25% ________________

27 Increase 36 by 80% ________________

28 Increase 150 g by 40% ________________

29 Increase $560 by 15% ________________

30 Increase 48 km by 30% ________________

31 Increase 65 kg by 12% ________________

32 Increase $90.50 by 66% ________________

33 Increase 124 by 6% ________________

34 A school has 460 students on its roll. If it grows by 5%, how many students will there be? ________________

35 Max is paid $25 per hour for gardening. If he is given an 8% increase, what will his new pay rate be? __________ per hour

36 A car dealer buys a car for $35 000. He must pay the government 15% tax on this amount before he sells it. How much will the car have cost him altogether? $________________

37 Toby bought a car using hire purchase. The price of the car was $5600. He has to pay for the car within a year, plus interest at 18%. How much will the car cost him altogether? $________________

ISBN: 9780170477710

N

Decreasing by a percentage

- Once again, there are several ways of doing this.

Example: Decrease 130 by 25%. — This means that we need to find 130 **minus** 25% of 130.

Decreased amount = 130 – 25% of 130
= 130 – 32.5
= 97.5

or

100% – 25% = 75%

Decreased amount = 130 x 75%
= 130 x 0.75
= 97.5

Otherwise use your calculator.

130  75 

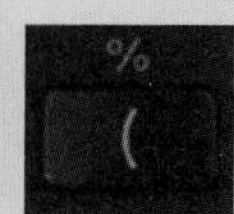

Calculate the following.

38 Decrease 84 by 20%

____

39 Decrease 256 by 35%

____

40 Decrease 240 g by 10%

____

41 Decrease $580 by 15%

____

42 Decrease 85 km by 12%

____

43 Decrease 124 kg by 60%

____

44 Decrease $435 by 55%

____

45 Decrease 116 cm by 5%

____

46 The value of a $545 000 house decreases by 4%. What is its decreased value? $__________

47 M Mart advertises '25% off everything in store'. Calculate these sale prices.

a A toaster which is normally $39.96. $__________

b A mattress which is normally $799.00. $__________

48 It is estimated that the value of an older car drops 15% each year. Estimate the value of Toby's $5600 car a year after he bought it. $__________

ISBN: 9780170477710

Rates

- Rates are often written using the word **'per'**.
- Rates are values with respect to a different unit.
- Examples: the amount paid per hour for a job;
  the cost per kilogram of oranges;
  the speed of a vehicle in kilometres per hour.

Examples:

1 4 kg of apples cost \$14.72. Calculate the cost of 3 kg of apples.

Method 1

Step 1: Write the information you are given in a brief statement.

4 kg cost **\$14.72**

Because you are trying to find a price, put the **price** on the **right**.

Step 2: Underneath write a parallel statement about what you want to know.

4 kg cost **\$14.72**

copy ↓

3 kg cost **\$14.72** $\times \frac{3}{4}$ = **\$11.04**

Multiply by either $\frac{3}{4}$ or $\frac{4}{3}$. Use $\frac{3}{4}$ here because 3 kg will cost **less**.

Or:

Method 2

Step 1: Work out how much 1 kg costs.

1 kg of apples costs **\$14.72** ÷ 4 = \$3.68

Step 2: Multiply the cost of 1 kg by how many you need.

So 3 kg of apples cost \$3.68 x 3 = **\$11.04**

2 Laundry powder comes in two different sizes. Which is better value for money?

Earthwear

\$16.94

1.8L

Squeaky Clean

\$37.56

4L

Step 1: Calculate the price per litre.

\$16.94 ÷ 1.8 = \$9.41   \$37.56 ÷ 4 = \$9.39

Step 2: Identify the product that is cheaper per litre.

Squeaky Clean is cheaper at \$9.39 per litre.

ISBN: 9780170477710

N

Answer the following questions.

1 It costs $13.50 for 3 kg of kūmara. Calculate the cost of 5 kg of kūmara.

____

____

2 It costs $6.84 for 6 m of fabric. Calculate the cost of 1.5 m of fabric.

____

____

3 **a** It costs NZ$1 to buy US$0.62. Amelia needed US$400 to take on a trip to the United States of America. To the nearest dollar, how much will it cost her in New Zealand dollars?

____

____

b She had some United States money left over when she returned to New Zealand. The exchange rate had remained the same. She received NZ$77.42 for the leftover money. How many United States dollars did she bring back?

____

____

Show your reasoning for the following calculations.

4 Calculate which size of toothpaste is better value for money.

A $3.23 — 170 g

B $2.25 — 90 g

____

____

5 Calculate which size of honey is better value for money:

a 400 g pot for $5.20 or a 2.5 kg tub for $29.95.

____

____

ISBN: 9780170477710

Ratios

- Ratios show how an amount is split into several shares, usually of different sizes.
- All shares must use the **same units**.
- Ratios use only **whole numbers**, not decimals or fractions.
- A **colon** (:) is used to separate the two shares. For example, 2:3 means one share made up of 2 parts and the other share made up of 3 parts.
- Ratios do **not** have units.

Example: $72 needs to be shared between two people in the ratio 3:5.

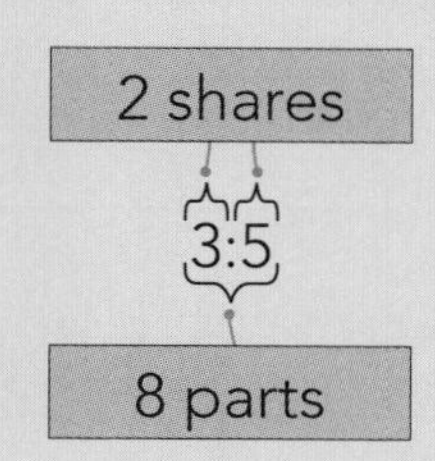

When there is a quantity that needs to be split using a ratio, follow these steps.

Step 1:	Find the total number of parts by adding the numbers in each share:	3 + 5 = 8
Step 2:	Find the value of one part by dividing the total by the number of parts:	$72 ÷ 8 = $9
Step 3:	Multiply the value of one part by the number of parts in each share:	(3 x 9):(5 x 9)
	So the ratio is	27:45.

That is, one person gets $27 and the other gets $45.

Another way to think of this:

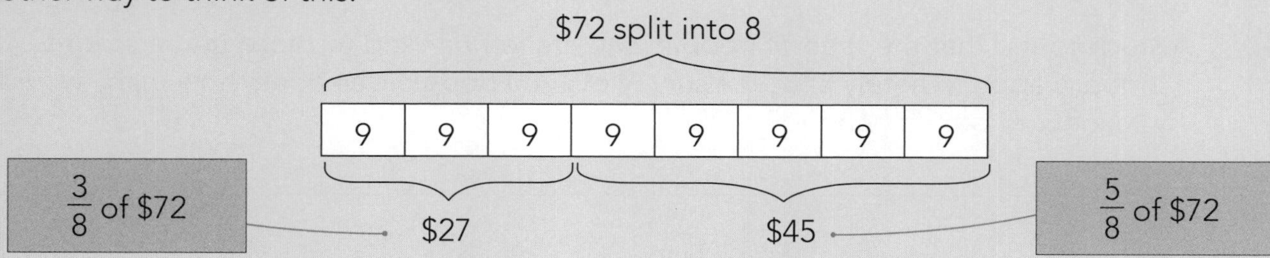

Note: Ratios don't have units or decimals, but your answer could have them.

Share the quantities in the given ratios.

1 $90 in the ratio 2:7

____

____

2 $55 in the ratio 5:6

____

____

3 240 g in the ratio 2:3

____

____

4 160 mL in the ratio 1:7

____

____

5 2100 m in the ratio 4:3

6 $92.25 in the ratio 3:6

7 A 15 kg bag contains apples and oranges in a ratio of 3:2. How many kilograms of apples are in the bag?

8 The ratio of students who play a sport to those who don't is 3:1. If there are 552 students on the school roll, how many play a sport?

9 Two friends were paid to paint a fence. Kahu painted for 3 hours, but Leah painted for only 2 hours. In total, they were paid $130. To be fair, how should they split the money?

10 A study found that the ratio of people who are left handed to those that are right handed is approximately 3:17. If a survey of 1280 people was done, how many would you expect to be left handed?

11 The ratio of doctors to nurses in an emergency room is 1:5. If there are 36 staff working, how many of them are doctors?

12 The ratio of students in Pia's class who bus to those that don't bus to school is 5:2. If there are 28 students in Pia's class, how many of them bus?

ISBN: 9780170477710

Calculations with money

- Remember, money calculations are the same as those for decimals.
- Sometimes you will need to **round** your answers.
- Remember, never round until **after** you have completed your calculations.
- While paying electronically may appear slightly more expensive, there may be other benefits such as convenience and loyalty rewards which affect your decision.

Money rounding in New Zealand

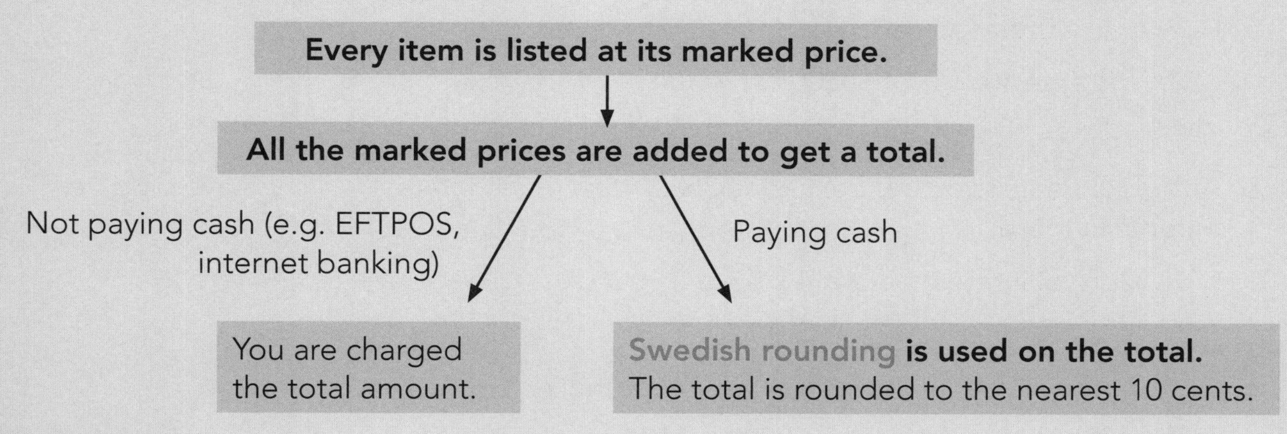

Example:

Total price (electronic price)	Swedish rounding (cash)
$13.98	$14.00
$5.47	$5.50
$329.32	$329.30
$19.03	$19.00

Cheaper to pay electronically (i.e. EFTPOS). ($13.98, $5.47)

Cheaper to pay cash. ($329.32, $19.03)

1 Complete the table.

Total price (electronic price)	Swedish rounding (cash)	Which is cheaper?
$47.38	$47.40	Electronic
$5.32		
$0.47		
$234.54		
$1290.72		
$23.69		
$315.98		

ISBN: 9780170477710

N

Simple interest

- When you deposit money into the bank, the bank pays you 'interest' for the use of that money.
- When you borrow money, you have to pay 'interest' because you are using the bank's money.
- Simple interest: the **same amount of interest** is paid at every interval (usually each year).

Examples:

1 Huia has $2000 and she borrows another $3500 off her parents in order to buy a car for $5500. Huia will pay them 5% interest on the $3500 each year.

a How much interest does she pay them each year?

Interest = 5% of $3500
= $175

b After four years she pays back the $3500. How much interest will she have paid her parents in total?

Total interest paid = 4 x $175
= $700

c How much will the car have cost her altogether?

Total cost = $2000 + $3500 + $700
= $6200

2 Jacob is saving in order to buy an electric bike. He puts $4500 into a deposit account which will pay him 2.5% interest each year.

a How much interest does he earn each year?

Interest earned = 2.5% of $4500
= $112.50

b He leaves this money in the bank for two years. How much interest will he have earned in total?

Total interest earned = 2 x $112.50
= $225

Answer the following questions. All interest rates are annual.

2 Complete the table.

Value of loan	Rate of interest	Interest per year	Number of years	Total value of interest
$2000	2%	$40	6	$240
$500	4%		3	
$7000	3.5%		4	
$12 000	8.5%		7	

ISBN: 9780170477710

3 Calculate the simple interest earned each year when $5000 is invested at 3%. $________

4 Calculate the total simple interest earned when $2500 is invested at 6% for three years. $________

5 Calculate the total simple interest earned when $45 000 is invested at 4% for five years. $________

6 Calculate the total simple interest earned when $250 000 is invested at 5% for three years. $________

7 **a** Tane has lent his sister $12 000 for two years to help her set up a business. His sister will pay him 5% simple interest on the total amount of the loan each year. Calculate the interest he earns each year. $________

b How much interest will he have earned in total by the end of two years? $________

c Calculate the total amount that his sister will owe him. $________

d She repays the total she owes him in equal monthly amounts over the two years. How much will she pay him each month? $________

8 **a** Moana borrowed $2000 from her grandmother to buy a new computer. She agreed to pay her grandmother 4% interest each year. How much interest will she need to pay each year? $________

b How much interest will she have paid altogether after four years? $________

c If she pays her grandmother back after four years, how much will the computer have cost her in total? $________

9 **a** Katie worked during her holidays and saved $3000. Her dad said he would pay her 6% of $3000 per year if she didn't spend any of the $3000, or the interest for two years. How much interest would her dad pay her each year? $________

b If she doesn't spend any of the money, how much will she have altogether by the end of two years? $________

10 Which of the following will give the total value if $2000 is invested at 3% simple interest per year for four years?

A $2000 x 1.03 x 4

B 4($2000 x 0.3) + $2000

C $2000 + 4 x $2000 x 0.03

D $2000 + 4 x $2000 x 0.3

N

GST

- GST stands for **Goods and Services Tax**.
- This is a tax of 15% on all goods sold and on all services.
  - In shops, GST is added to the cost price of items and to the shop's profit margin (mark-up).
  - GST is added to the hourly rate, parts and travel time of plumbers, electricians, etc.

Examples:

1 A greengrocer buys broccoli at \$2.50 per head. He adds a \$1 profit margin.

Sale price = (\$2.50 + \$1) x 1.15 = \$4.025. Round to \$4.00.

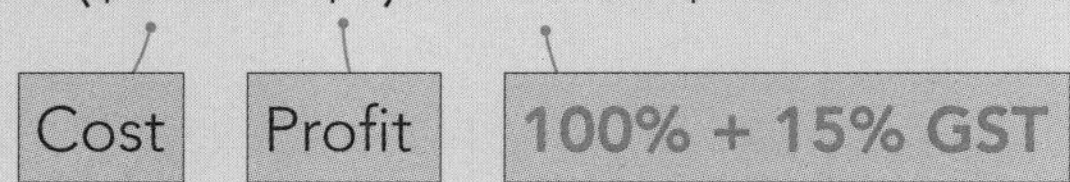

2 A builder's base rate (what he gets) is \$40 per hour.

a How much GST must be added to this rate? \$40 x 15% = \$6

b How much per hour should he charge his customers? \$40 + \$6 = \$46 per hour

Calculate how much GST would be charged on these amounts.

11 \$42 \$__________ **12** \$11 \$__________

13 \$513 \$__________ **14** \$1875 \$__________

Calculate the GST-inclusive prices for these amounts.

15 \$10 \$__________ **16** \$8.50 \$__________

17 \$79 \$__________ **18** \$362 \$__________

Answer the following questions.

19 Rob owns a butcher shop. The cost of ingredients and the time needed to make sausages come to \$7 per kilogram, and he adds on a \$2 per kilogram profit margin.

Calculate the selling price for his sausages after GST is added. \$__________

20 Miranda is an electrician. Her basic charge-out rate is \$38 per hour. Calculate the GST-inclusive rate that she should charge her customers. \$__________

21 A car on a dealer's yard cost him \$5000. He adds a \$2000 profit margin. How much GST will he need to add? \$__________

What will be its price on the yard? \$__________

ISBN: 9780170477710

Estimations/approximations

- Sometimes an exact answer isn't necessary, so you can just estimate the answer.
- In order to estimate, round every number to **one significant figure**, and then do the calculations required.
- Don't forget to use BEDMAS.

Where values are similar, round to one significant figure

- When rounding to one significant figure, there should be just one non-zero digit in the rounded number.
- Use the same rules as rounding to decimal places.

Locate the digit to the **right** of the last required digit.

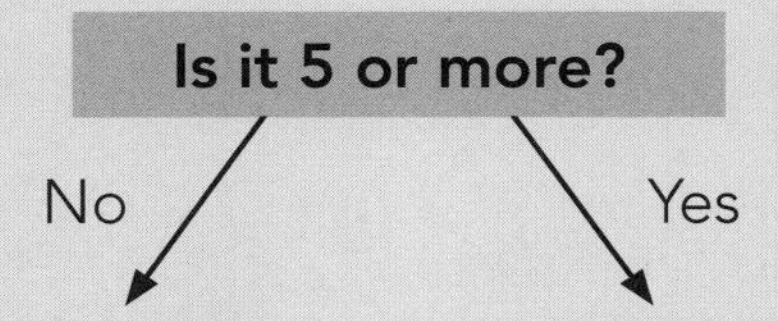

No	Yes
Replace it and everything after it with 0s.	Increase the previous digit by one and replace everything after it with 0s.

Examples:

1 Round \$9.70 to one significant figure.

Is \$9.70 closer to \$9.00 or \$10.00?

The 7 is more than 5 ⇒ increase the 9 to 10 and replace the 7 with a 0.

\$9.70 = \$10.00 (1 sf)

2 Round \$246 to one significant figure.

Is 246 closer to 200 or 300?

The 4 is less than 5 ⇒ leave the 2 alone and replace the 4 and 6 with zeros.

\$246 = \$200 (1 sf)

Notice that both the rounded amounts have just one non-zero digit at the start.

3 A list of prices is given. Estimate the total.

This symbol means 'is approximately equal to'.

Socks	\$14.62 ≈ \$10.00
Beanie	\$23.40 ≈ \$20.00
T-shirt	\$48.95 ≈ \$50.00

The approximate total is \$80.00.

10 + 20 + 50 = 80

Estimate the answers to the following. Do not use a calculator.

22 3.49 + 9.73 ≈ 3 + 10

= ______

23 1.19 + 5.52 + 3.41 ≈ ____ + ____ + ____

= ______

ISBN: 9780170477710

N

24 $38.03 - 17.57 \approx$ ______ $-$ ______

$=$ ______

25 $349 + 953 \approx$ ______ $+$ ______

$=$ ______

26 $4.2 + 9.1 \times 4 \approx$ ______ $+$ ______ $\times$ ______

$=$ ______

27 $6.2 \div 2.5 + 7.35 \approx$ ______ $\div$ ______ $+$ ______

$=$ ______

Where values are not similar

Step 1: Find a value near the middle of the range of values. Round that to 1 sf.

Step 2: Round all other values to the same number of decimal places.

Hints:
1. It sometimes gives a more accurate estimate if you round some figures up and others down.
2. When dividing, try to round so you get numbers that divide easily.

Examples:

1 A list of prices:

Glue	$8.59 ≈ $10.00
Nails	$22.90 ≈ $20.00
Timber	$182.05 ≈ $180.00

Middle value = $20 (1 sf). So round other values to the nearest $10.

The approximate total is $210.00.

10 + 20 + 180 = 210

2 Three friends were going away together for a weekend and will share the expenses. They spent $258.25 at the supermarket, $85.92 on fuel, and $16.00 on a bag of cherries.

Both of these round up, so round one down and the other up.

Estimation of the amount each will need to pay: $\frac{260 + 90 + 10}{3} = 120$

3 Five friends shared the $127.95 it cost them for a meal.

Estimation of the amount each will need to pay: $\frac{\$130}{5} = \$26$

28 $342 + 12 + 9 \approx$ ______ $+$ ______ $+$ ______

$=$ ______

29 $5488 + 634 - 115 \approx$ ______ $+$ ______ $-$ ______

$=$ ______

30 $42\,500 \div 7.045 \approx$ ______ $\div$ ______

$=$ ______

31 $37.4 \times 19 + 269 \approx$ ______ $\times$ ______ $+$ ______

$=$ ______

32 $\frac{11\,976}{361} \approx$ ______

$=$ ______

33 $1.8(23.45 - 8.112) \approx$ ______(______ $-$ ______)

$=$ ______

ISBN: 9780170477710

34 $1469 \div (685 - 355) \approx$ ____ $\div$ (____ $-$ ____)

$=$ ______

35 $(213 + 36.6) \div 5.2 \approx$ (____ $+$ ____) $\div$ ____

$=$ ______

36 Annie has $87.59 in her bank account. She went to a movie which cost her $8.50, bought a milkshake which cost her $4.30, and caught the bus home which cost her $1.90. Estimate the amount left in her bank account.

____

37 About how much change from a $50 note would you get if you bought two pairs of socks at $4.20 each and three T-shirts at $9.95 each?

____

38 Estimate the total and the change expected from a $50 note.

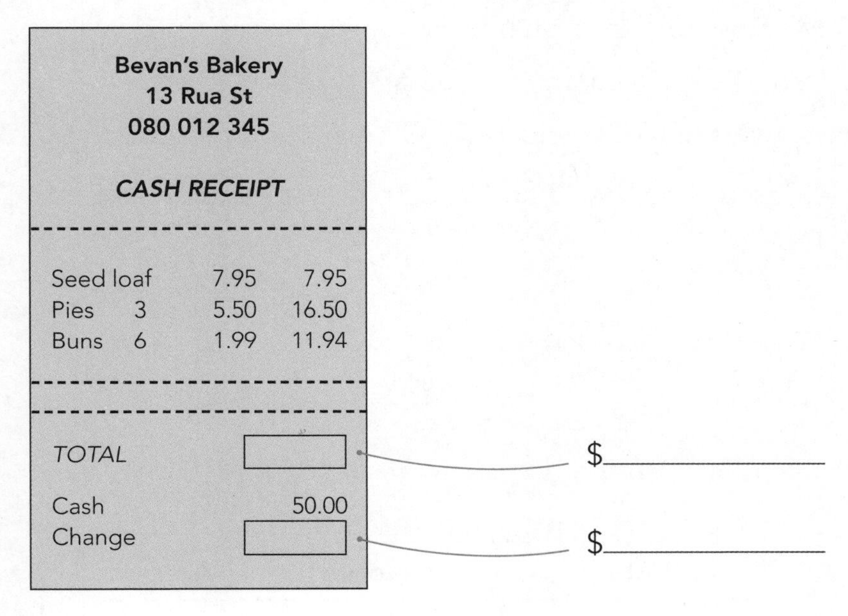

Bevan's Bakery
13 Rua St
080 012 345

CASH RECEIPT

Seed loaf		7.95	7.95
Pies	3	5.50	16.50
Buns	6	1.99	11.94

TOTAL $______

Cash 50.00

Change $______

39 Estimate the total and the change expected from a $100 note.

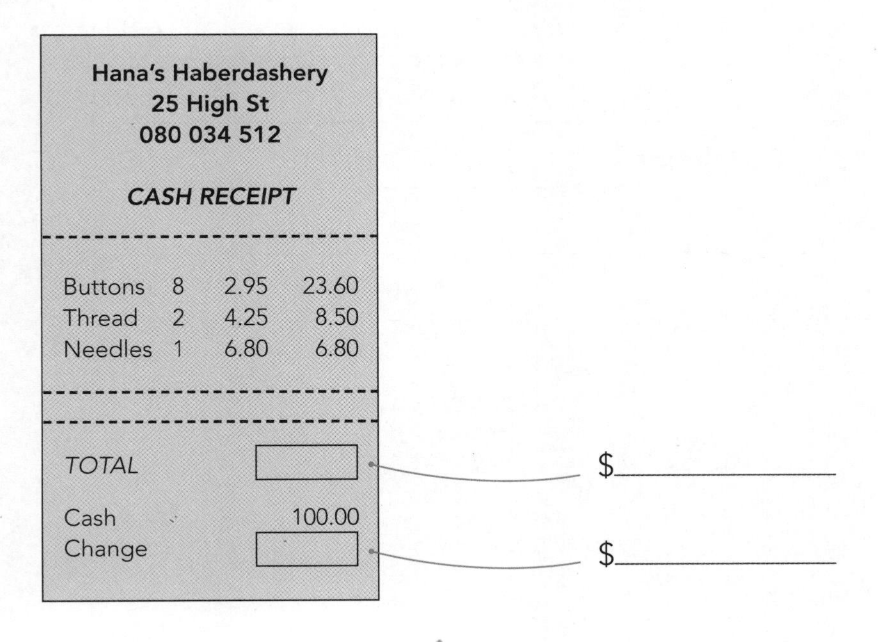

Hana's Haberdashery
25 High St
080 034 512

CASH RECEIPT

Buttons	8	2.95	23.60
Thread	2	4.25	8.50
Needles	1	6.80	6.80

TOTAL $______

Cash 100.00

Change $______

ISBN: 9780170477710

N

Payslips

- These show the detail of how your take-home pay is calculated.

You need to know how to calculate the following.

1 **Time and a half:** this is the pay rate for working on public holidays or for overtime.

Time and a half = your normal rate + half your normal rate.

For example: Penny is paid $21.50 per hour. She works for five hours on Matariki.

She will be paid $21.50 x **1.5** = $32.25 per hour.

For working five hours on Matariki, her gross pay = 5 x $32.25

= $161.25

2 **PAYE tax:** this is the tax that must be paid on earnings.

The rate of PAYE tax depends on how much you earn:

- Less than $14 000 per year means you are taxed at 10.5%.
- Earnings between $14 001 and $48 000 means you are taxed at 17.5%, etc.

For example: Tama earns less than $14 000 per year and he earned $456.35 last week. He will have to pay $456.36 x 10.5% = $47.92 tax.

Rounded from $47.9178

Example:

Employees are often paid every two weeks.

Income	Rate	Hours	Amount
Regular	27.50	80	$2200.00
Time and a half	41.25	5.5	$226.88
		Total gross pay	$2426.88
	Deductions	PAYE (10.5%)	$254.82
		Net pay	$2172.06

41.25 x 5.5 = 226.**875**. This must be rounded to the nearest cent.

Gross pay = Regular pay + Time and a half

PAYE tax rate: 10.5% (rounded from $254.8224)

Net pay = Total gross pay – PAYE tax

ISBN: 9780170477710

Calculate time-and-a-half pay for each of the following normal hourly rates. Round to the nearest cent.

40 $23.00 $__________ **41** $25.50 $__________

42 $27.25 $__________ **43** $32.50 $__________

Calculate the PAYE tax (10.5%) that will have to be paid on the following. Round to the nearest cent.

44 $480.00 $__________ **45** $342.00 $__________

46 $875.50 $__________ **47** $1084.25 $__________

Fill in the grey boxes and round values to the nearest cent.

48

Income	**Rate**	**Hours**	**Amount**
Regular	21.00	40	$
Time and a half		7	$
		Total gross pay	$
	Deductions	PAYE (10.5%)	$
		Net pay	$

49

Income	**Rate**	**Hours**	**Amount**
Regular	27.50	80	$
Time and a half		13	$
		Total gross pay	$
	Deductions	PAYE (10.5%)	$
		Net pay	$

50

Income	**Rate**	**Hours**	**Amount**
Regular	38.00	80	$
Time and a half		5.5	$
		Total gross pay	$
	Deductions	PAYE (10.5%)	$
		Net pay	$

ISBN: 9780170477710

Bank statements

- You need to be able to understand how to read a bank statement.

Credit means money is put into the account.

Debit means is taken out of the account.

Balance is the amount of money left in the account.

Date	Transaction	Credit	Debit	Balance
14 Jan 22				$762.53
14 Jan 22	Cafe		$14.95	$747.58
15 Jan 22	Pay	$452.50		$1200.08
19 Jan 22	Rent		$120.00	$1080.08
22 Jan 22	Supermarket		$164.99	$915.09

$762.53 – $14.95 = $747.58

$747.58 + $452.50 = $1200.08

- Be careful. Sometimes bank statements appear with the latest data at the top, and sometimes it is at the bottom. Check the dates.

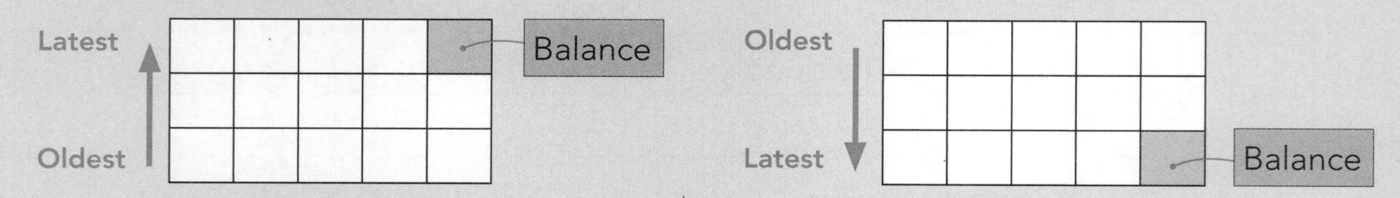

Answer the following questions. For each question make sure you check the dates to find out where to start.

51 Show the calculation that produced the missing values and add them to the table.

a a = ____________________ **b** b = ____________________

Date	Transaction	Credit	Debit	Balance
13 Jun 22	Takeaways		$35.76	$820.81
14 Jun 22	Supermarket		$54.75	a
15 Jun 22	Pay	$598.40		b
15 Jun 22	Dairy		$12.49	$1351.97

52 Show the calculation that produced the missing values and add them to the table.

a a = ____________________ **b** b = ____________________

Date	Transaction	Credit	Debit	Balance
5 Aug 22	Pay	$348.90		b
3 Aug 22	Fuel		$61.46	$119.25
30 Jul 22	Supermarket		$39.95	a
26 Jul 22	Canteen		$8.75	$220.66

ISBN: 9780170477710

53 Fill in the grey boxes with the missing values.

Date	Transaction	Credit	Debit	Balance
6 Dec 22	Pay	$675.00		$698.23
8 Dec 22	Bike repair		$45.00	$653.23
8 Dec 22	Rent		$350.00	$303.23
8 Dec 22	Supermarket		$105.62	$197.61
11 Dec 22	B Mart		$96.89	$
15 Dec 22	Babysitting	$80.00		$
16 Dec 22	Dairy		$13.25	$

54 Here is Tania's latest bank statement.

Date	Transaction	Credit	Debit	Balance
30 Nov 22	WOF		$72.00	–$45.65
29 Nov 22	Car repair		$166.00	$26.35
23 Nov 22	Supermarket		$156.55	$192.35
21 Nov 22	Rent		$255.00	$348.90
21 Nov 22	Pay	$598.70		$603.90

Why does she have a balance of –$45.65 on 30 November? ________

A She has $45.65 in bills to pay.

B The bank charged a fee of $45.65.

C There is $45.65 left in her account.

D She owes the bank $45.65.

55 Fill in the grey boxes with the missing values.

Date	Transaction	Credit	Debit	Balance
3 Nov 22	B Mart		$56.80	$32.55
5 Nov 22	Auto Supplies		$125.66	$
8 Nov 22	Cafe		$18.50	$
8 Nov 22	Pay	$328.50		$
9 Nov 22	Supermarket		$85.36	$

ISBN: 9780170477710

Mixing it up

1 **a** When this question was written, the population of New Zealand was 4 920 648. Write this number using words.

____

____

b Round 4 920 648 to the nearest thousand. ____________

c If 20% of the population is under 15 years old, about how many New Zealanders are under 15 years old? Round your answer to the nearest ten. ____________

2 **a** At Scott Base in Antarctica during December 2022, the average daily maximum temperature was 0.9°C and the average daily minimum was –3.8°C. Calculate the difference between these temperatures.

Difference = ________°C

b The total land area of Antarctica is 14.2 million km^2; 98% of this is covered by ice. Calculate the area covered by ice.

________ km^2

c At the end of the 2021 winter, the area covered by sea ice was 19 million square kilometres. By the end of the following summer (February 2022), this had decreased by 90%. Calculate the area of sea ice at the end of the summer.

________ km^2

d The number of emperor penguins on Antarctica in 2022 was estimated to be 6×10^5. Write this number without using powers.

____

e A scientist weighed an emperor penguin egg. His scale read 489.51 g. Round this mass to the nearest gram.

____________ g

f When a female emperor penguin lays an egg, she passes it to the male who keeps it warm for nine weeks. During this time he doesn't eat and loses 45% of his body mass. If his mass was 40 kg to start with, what would it be at the end of this period?

________ kg

g The cheapest trip to Antarctica costs US$12 500. It costs NZ$1 to buy US$0.64. Calculate the price of the trip in New Zealand dollars. Round your answer to the nearest five hundred dollars.

NZ$____________

 ISBN: 9780170477710

3 **a** Nikora works for his local fruit and vegetable shop. He is paid $23 per hour and time and a half when he works more than 40 hours in a week. This is his payslip for last week. Complete the payslip.

Income	Rate	Hours	Amount
Regular	23.00	40	$
Time and a half		5	$
		Total gross pay	$
	Deductions	PAYE (10.5%)	$
		Net pay	$977.79

b Here is a copy of his latest bank statement. Complete the statement.

Date	Transaction	Credit	Debit	Balance
5 Nov 22	Shoe Mart		$125.00	$230.69
7 Nov 22	Takeaways		$32.50	$
11 Nov 22	Dairy		$13.55	$
11 Nov 22	Pay	$977.79		$
15 Nov 22	Rugby club		$	$1087.43

c Nikora wants to buy a car which costs $3500. He has saved $2000, and his mum says she will lend him the extra $1500 that he needs, but he will have to pay her 5% interest each year.

i How much interest will he pay her each year? $__________

ii If he pays her back after three years, how much will the car have cost him **altogether**? $__________

iii He estimates that the running costs for his car will be:

Fuel	$60 per week
Insurance	$30 per month
Registration	$103 per year
Maintenance	$700 per year

Estimate the total running costs of his car for one year. $__________

ISBN: 9780170477710

Mathematical relationships

Reading axes

- **Axes** are the horizontal and vertical lines that border a graph.
- The **vertical** axis is the y-axis, the **horizontal** is the x-axis.
- The **origin** is the point (0, 0)

We speak of **one** axis or **several** axes.

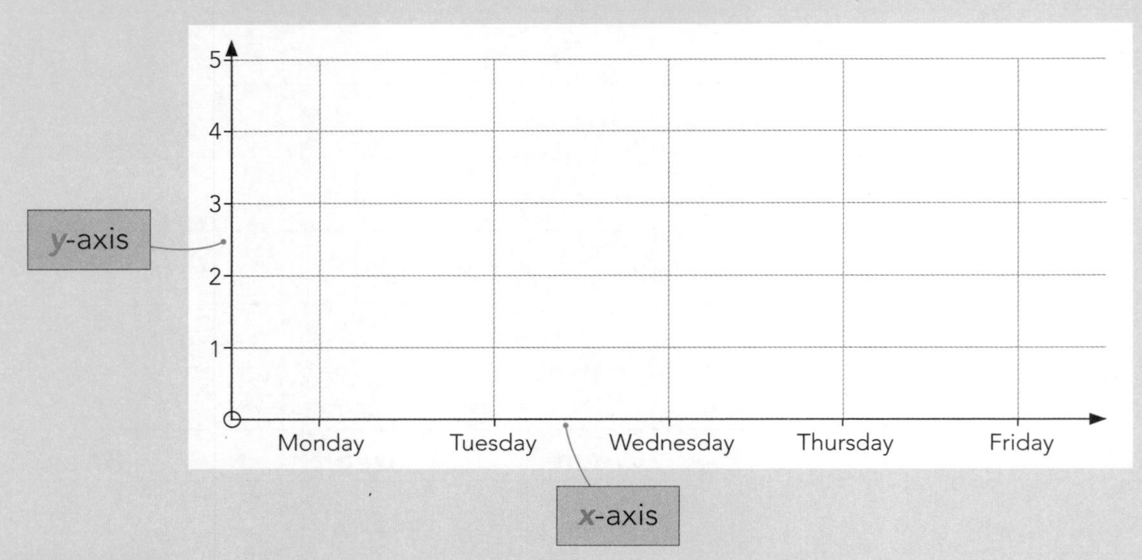

Examples:

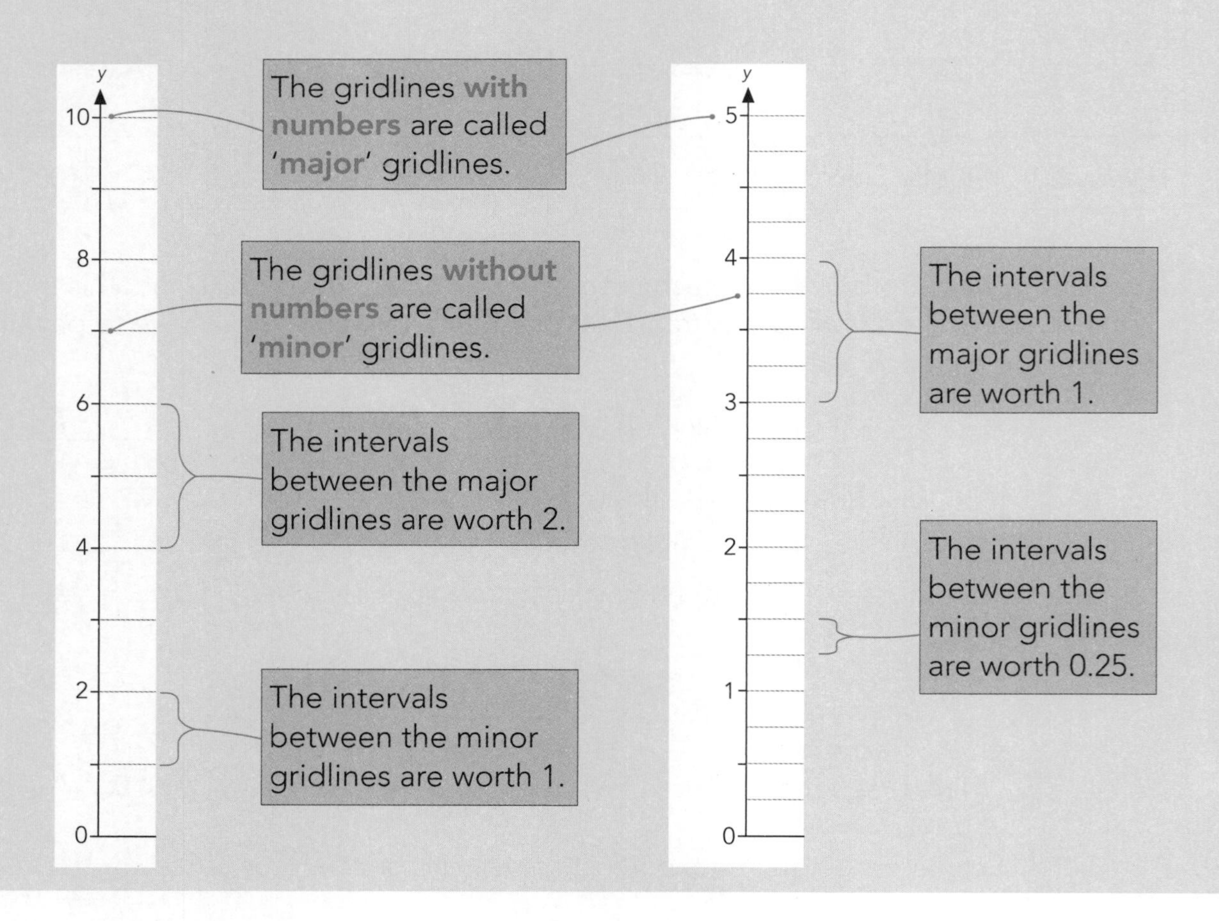

 ISBN: 9780170477710

Write the missing values from these axes.

1 Major ________ Minor ________

2 Major ________ Minor ________

3 Major ________ Minor ________

4 Major ________ Minor ________

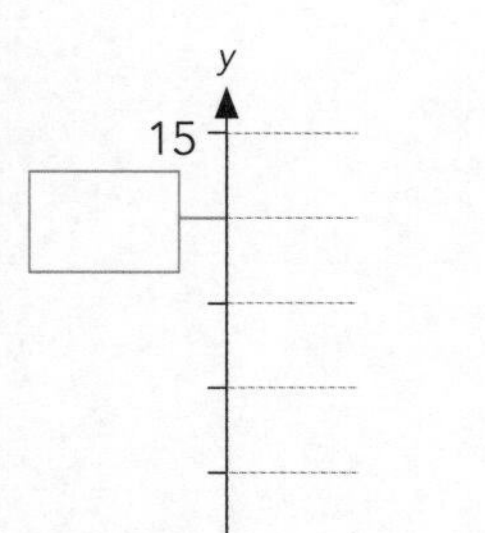
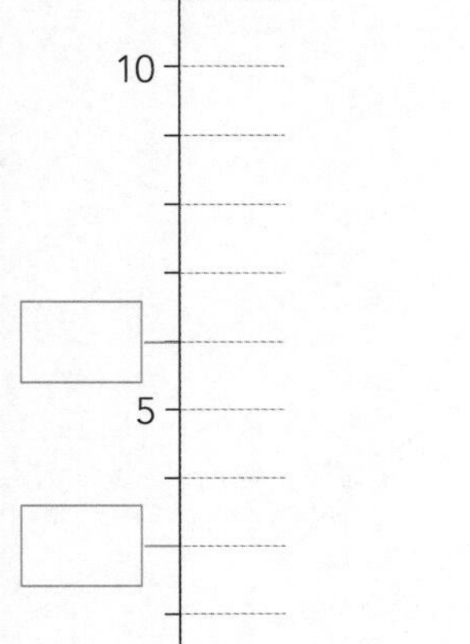

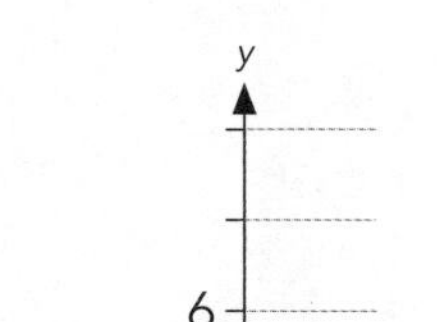
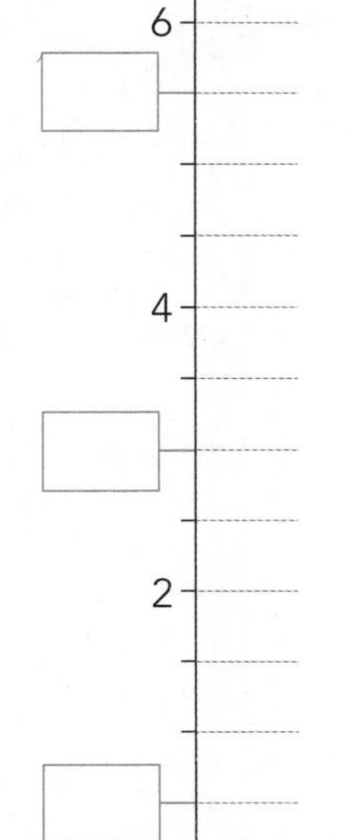

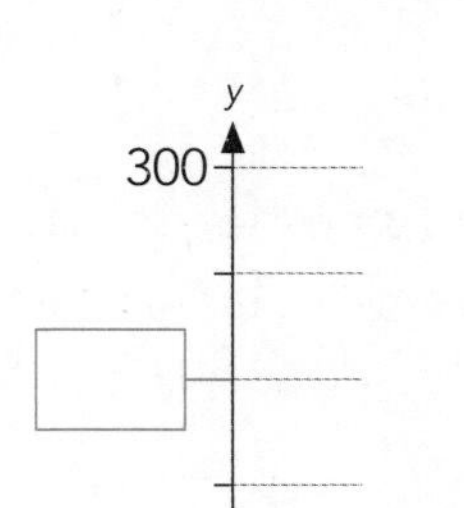
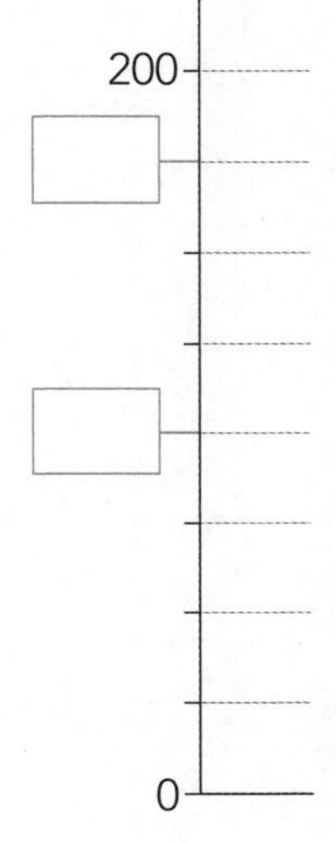

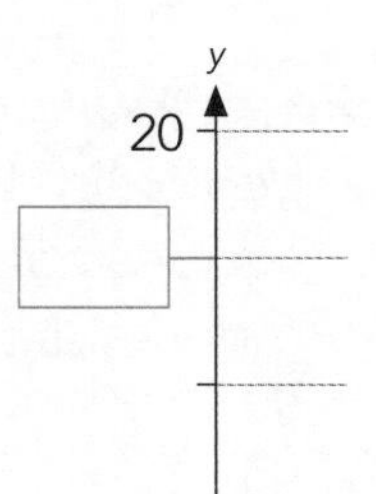
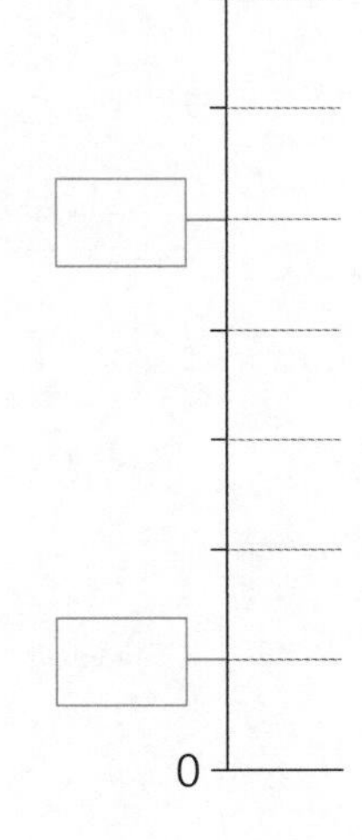

5 Major ________ Minor ________

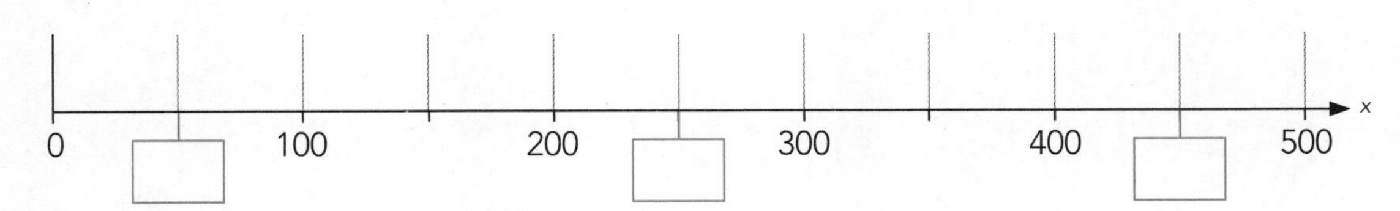

6 Major ________ Minor ________

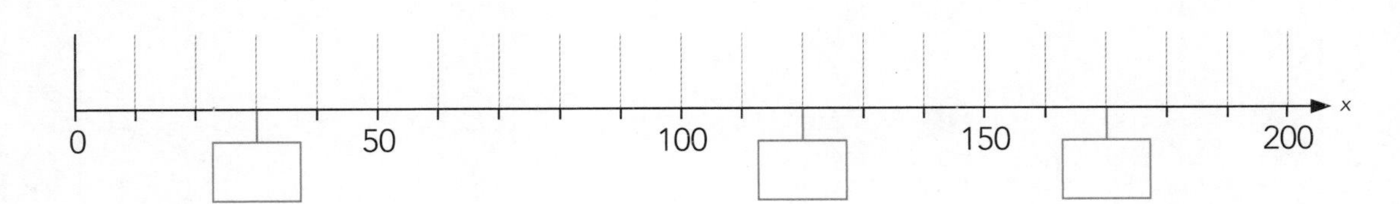

7 Major ________ Minor ________

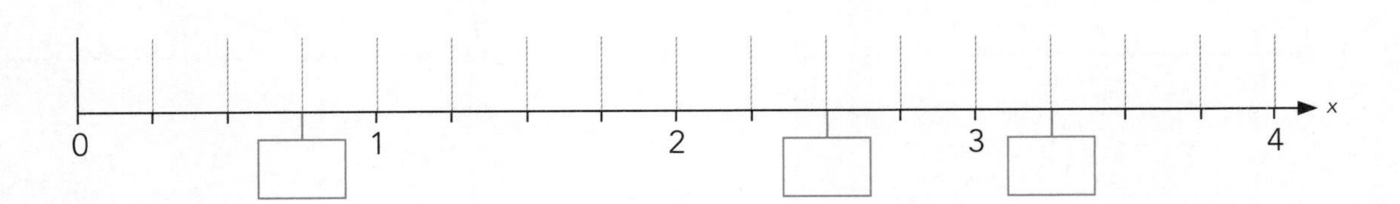

ISBN: 9780170477710

N

Coordinates

- The points on a graph can show the **relationship** between **two variables**.
- Each point represents **two** pieces of data about **one** object, e.g. the height and mass of a person.

Coordinate revision

- The *x* coordinate tells you how far to move to the **right**.
- The *y* coordinate tells you how far to move **up**.
- Coordinates are written in brackets, in alphabetical order: **(*x*, *y*)**

Examples:

1

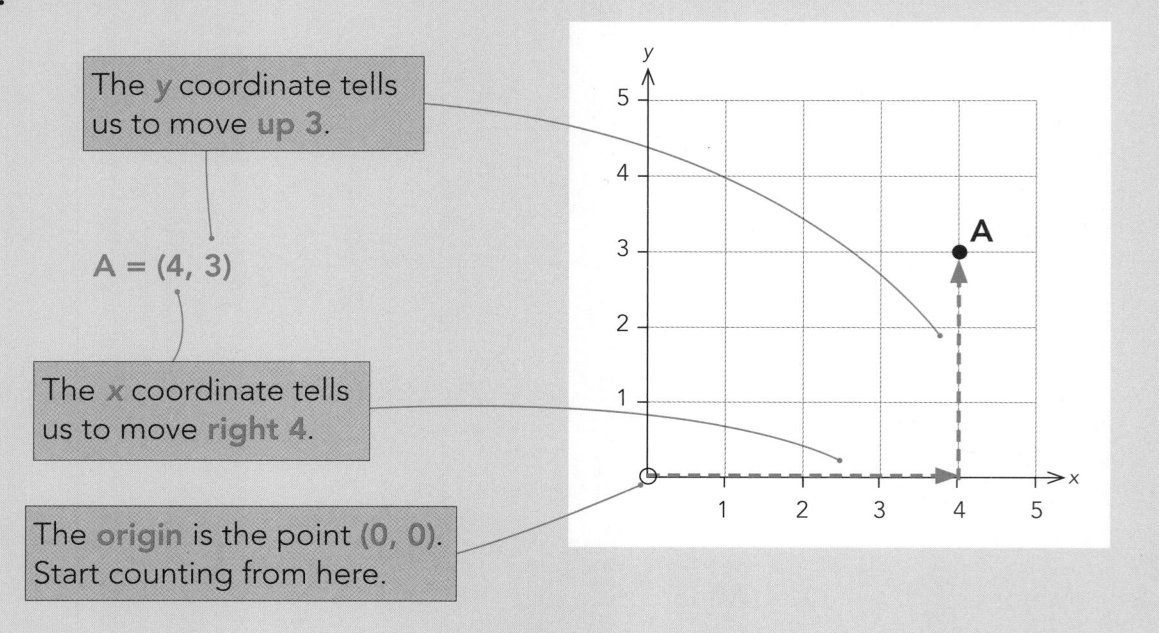

2 Coordinate B: (7, 5)

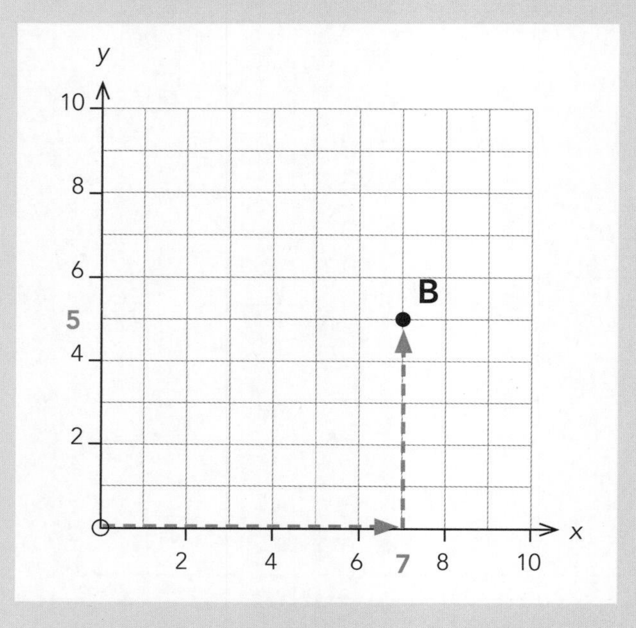

3 Coordinate C: (17, 19)

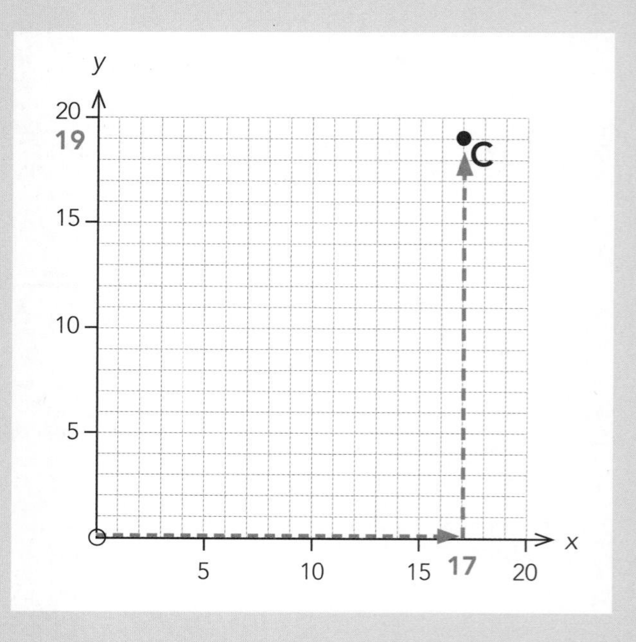

ISBN: 9780170477710

1 **a** Add the missing numbers to the axes and then write the coordinates for the points A, B, C and D.

A (______, ______) B (______, ______)

C (______, ______) D (______, ______)

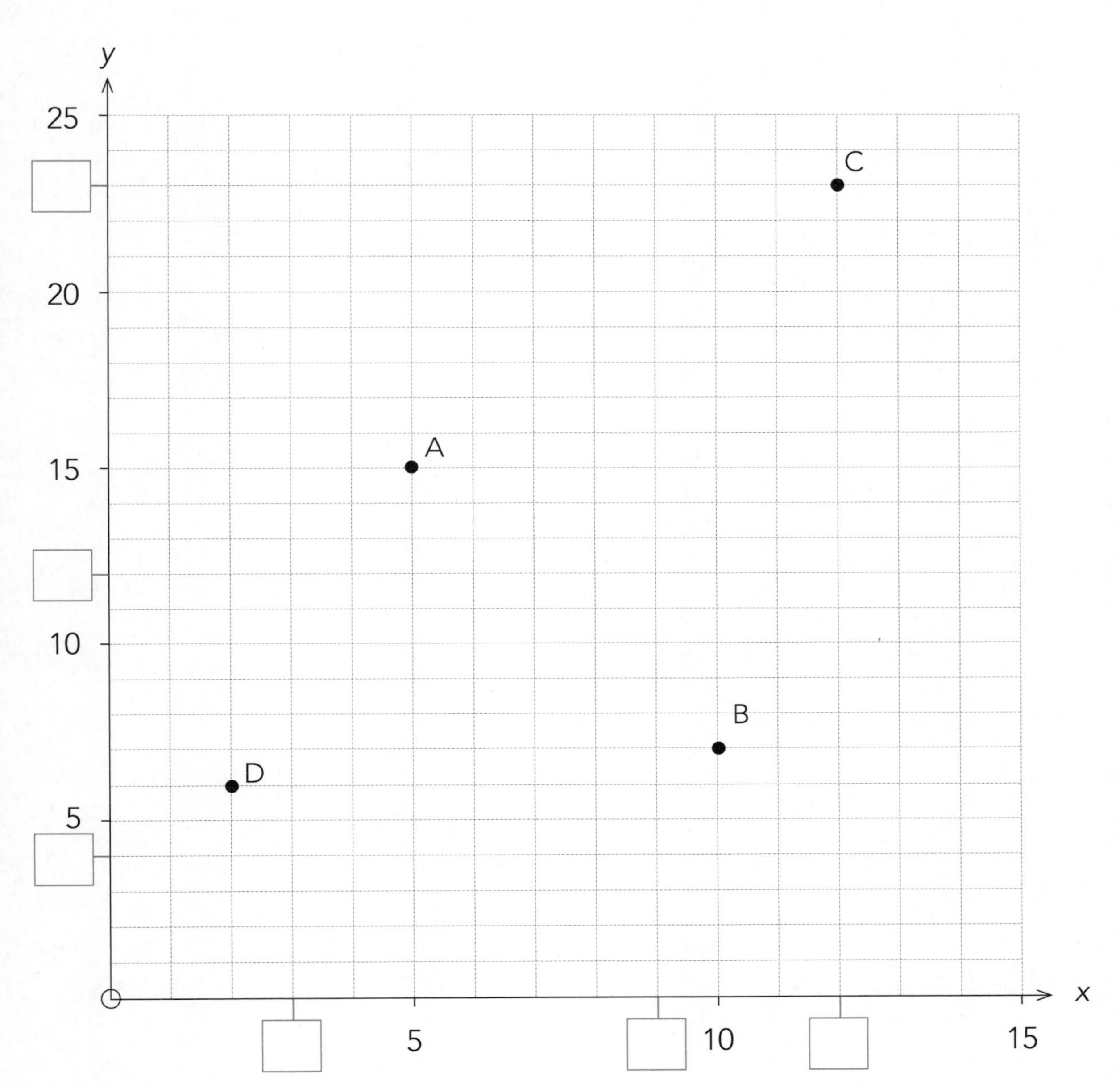

b Plot and label these points on the axes above.

E (8, 20) F (15, 11) G (1, 21) H (7, 0)

ISBN: 9780170477710

2 a Add the missing numbers to the axes and then write the coordinates for the points A, B, C and D.

A (______, ______) B (______, ______)

C (______, ______) D (______, ______)

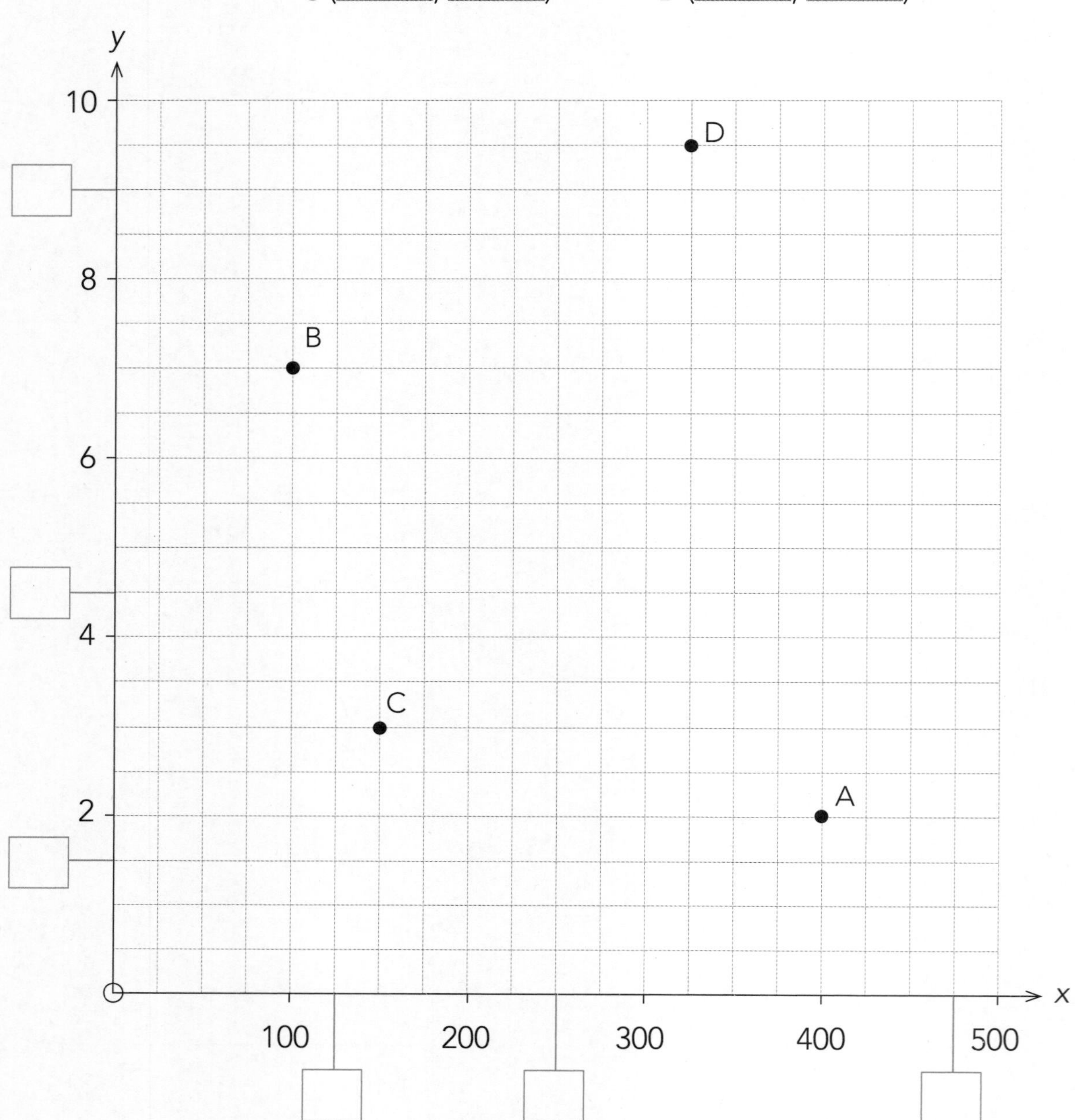

b Plot these points on the axes above.

x	***y***
50	2
350	6.50
450	8.50
75	0.5

ISBN: 9780170477710

Linear relationships

- When plotted, a linear relationship forms a **straight line** on a graph.
- An arrow at the end of the line tells you that the relationship keeps going.
- Reading coordinates of points on a line can help you to answer questions.

Reading coordinates

Example:

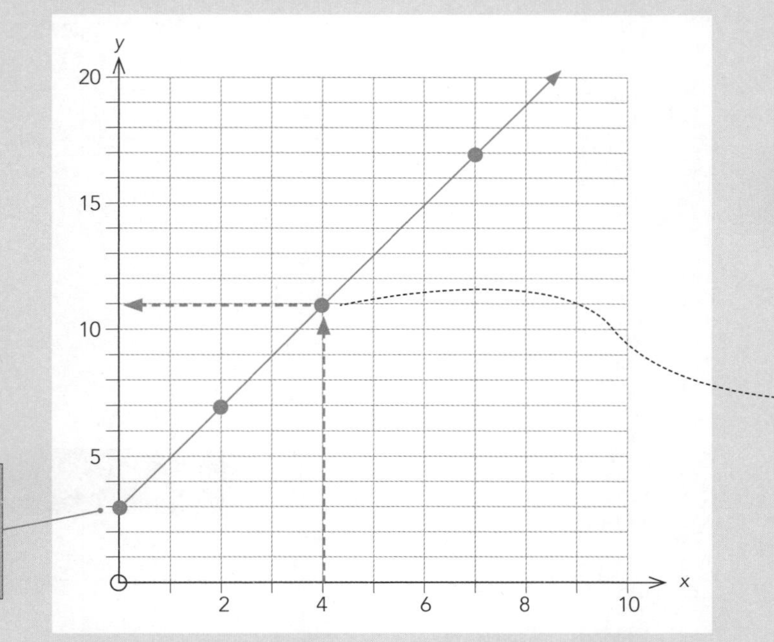

We call this point the **y intercept**.

x	y
0	3
2	7
4	**11**
7	17

Complete the tables of coordinates for these graphs.

1

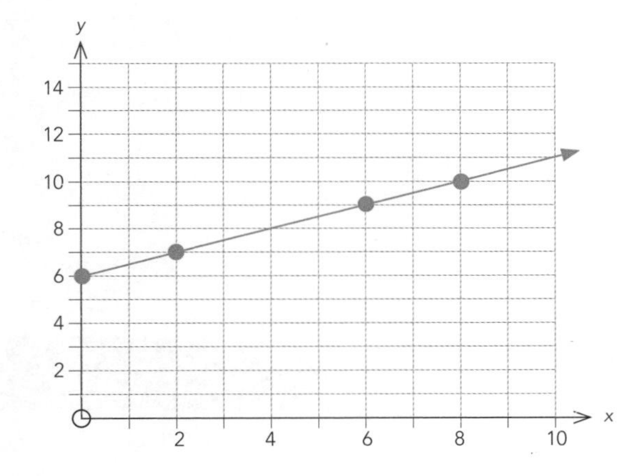

x	y
0	
2	
6	
8	

2

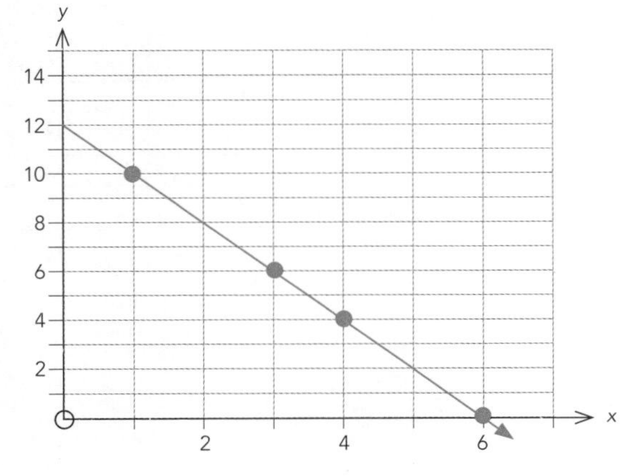

x	y
1	
3	
4	
6	

ISBN: 9780170477710

N

3

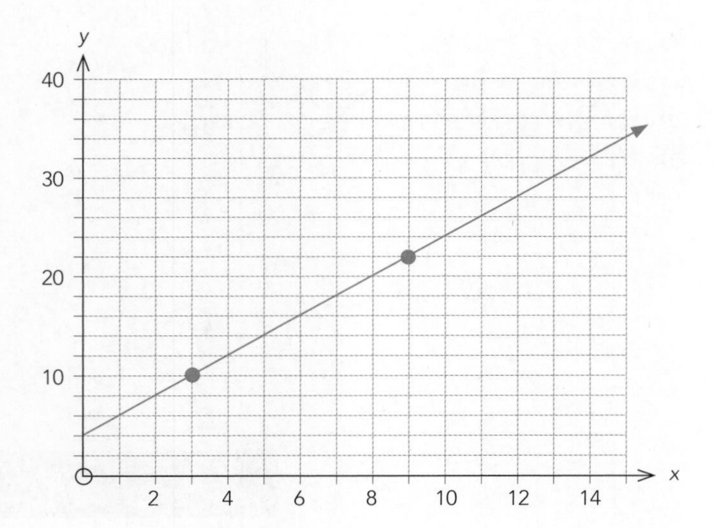

x	y
3	
9	
	30
	34

4

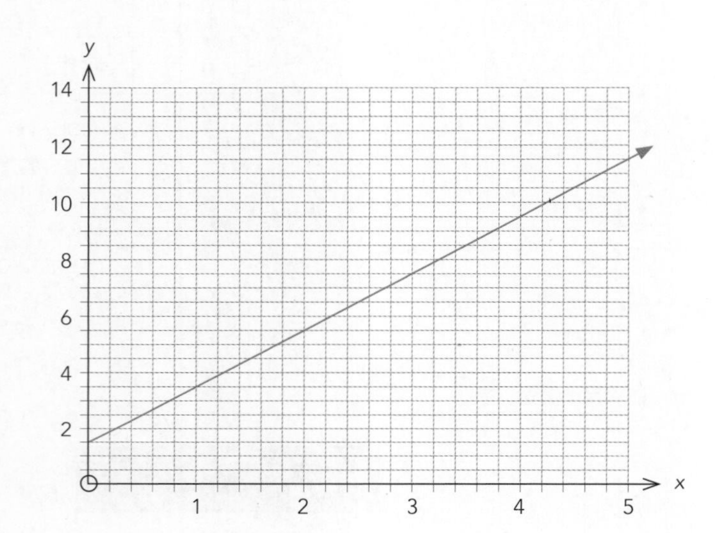

x	y
0	
2	
4	
5	

5

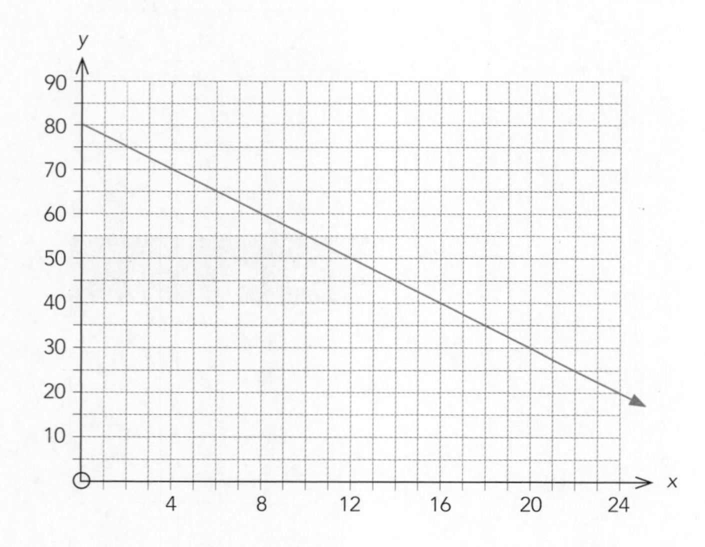

x	y
0	
	45
	35
24	

ISBN: 9780170477710

Plotting points and drawing straight lines

- When plotted, a linear relationship forms a **straight line** on a graph.
- Before you rule the line, make sure you have plotted **at least three points**.

Plot points from coordinates or a table.

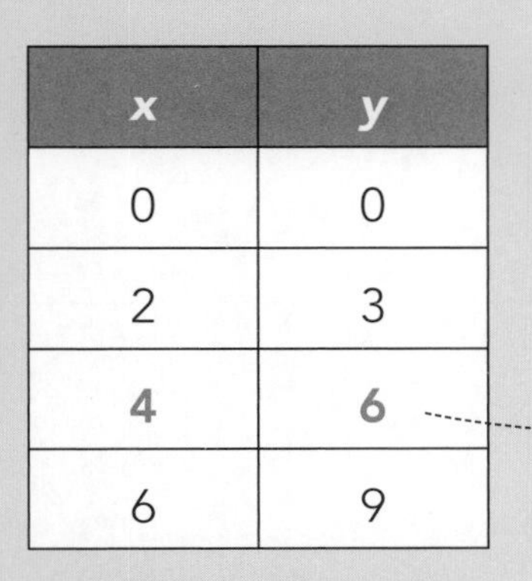

x	y
0	0
2	3
4	**6**
6	9

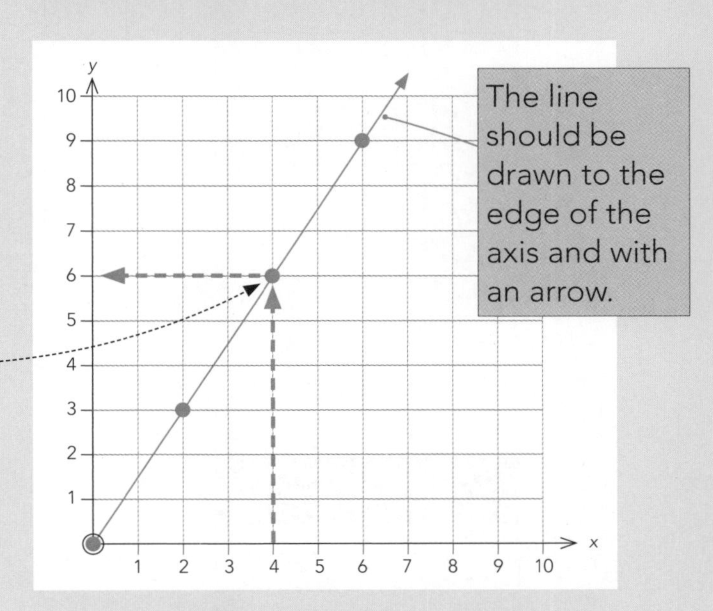

Plot these points on the graph, then join them with a straight line.

6

x	y
0	0
3	2
6	4
9	6

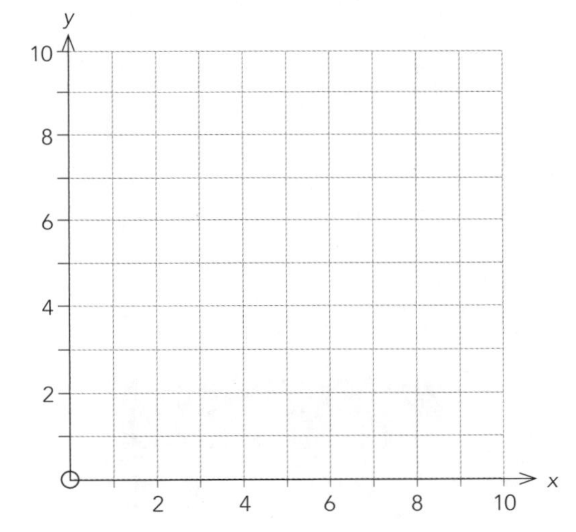

7

x	y
2	1
4	2
6	3
8	4

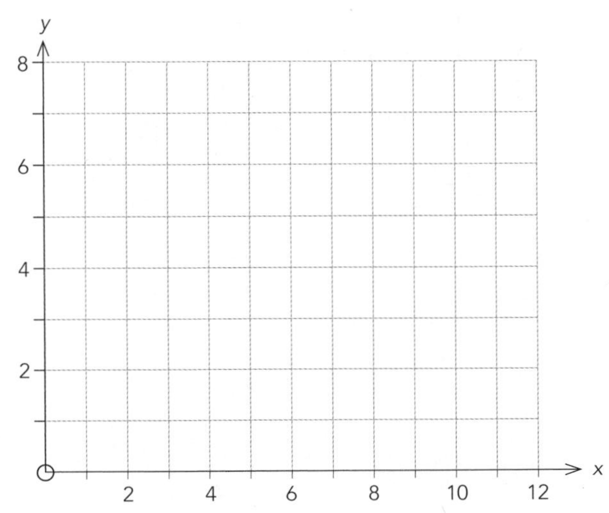

ISBN: 9780170477710

N

8

x	y
4	3
8	6
12	9
16	12

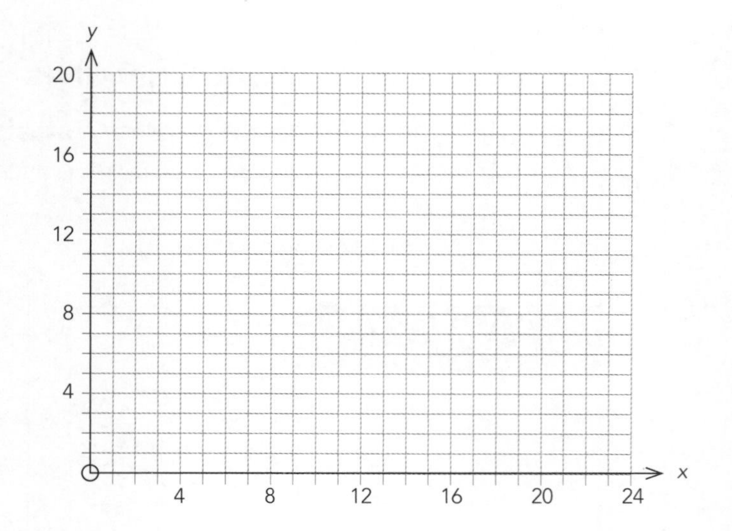

9

x	y
6	24
18	16
30	8
42	0

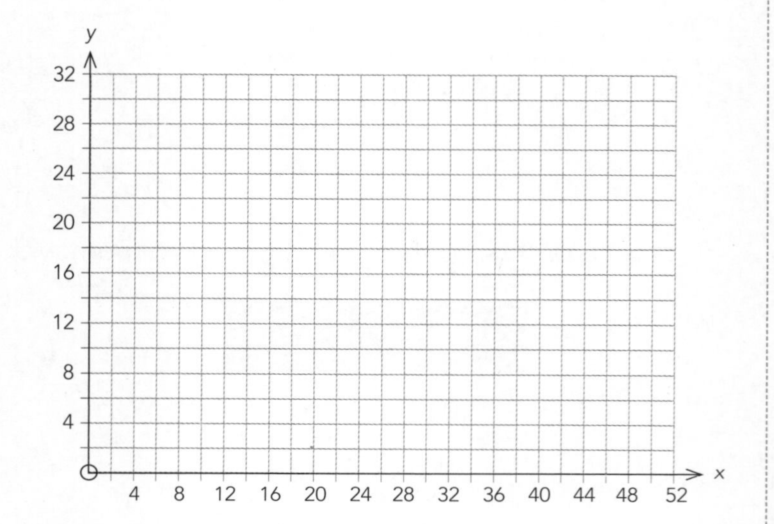

10

x	y
2	1.2
4	2.4
6	3.6
8	4.8

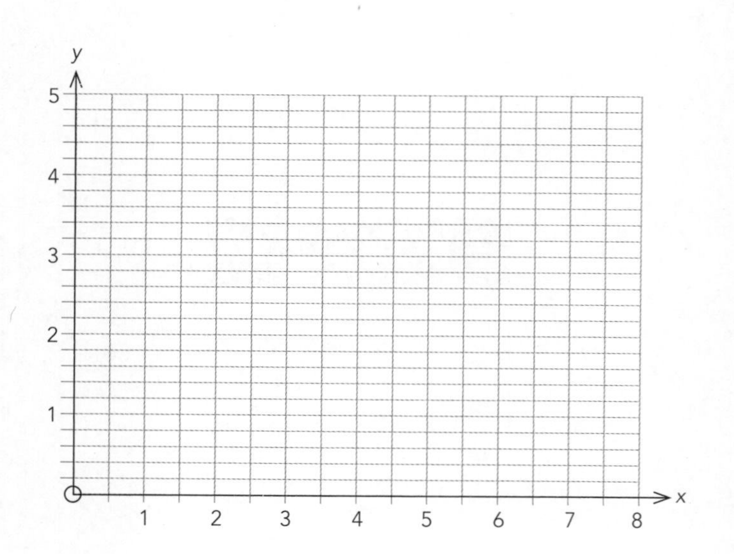

ISBN: 9780170477710

Applications

- Graphs can be very helpful when solving a problem.

Understanding graphs

- There are **three** things that we need to understand about linear graphs:

1 Where it passes through the y-axis: the **y intercept**.

This is the value for y when **$x = 0$**.

In this case, the y intercept is 3.

The point (0, 0) is called the **origin**.

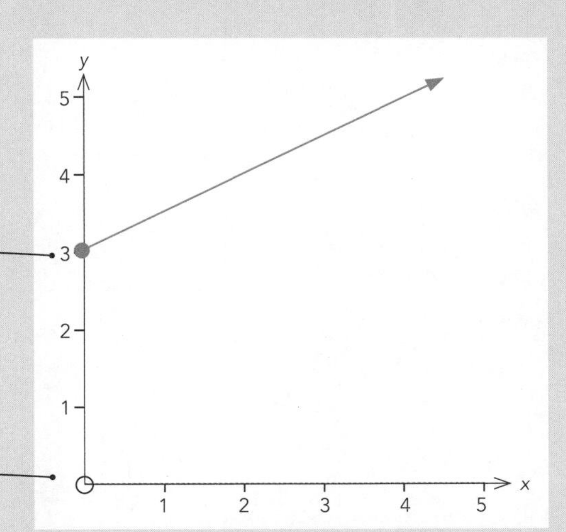

2 How **steep** the line is: the size of the **gradient**.

This tells us how fast the y values change when x changes.

Steep gradient means y values change **quicker**.

Shallow gradient means y values change more **slowly**.

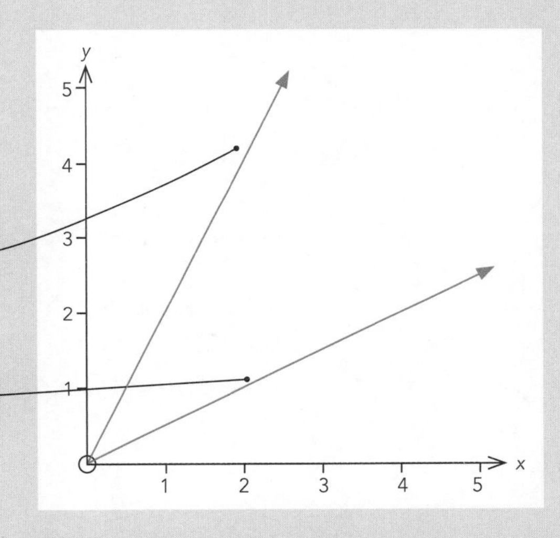

3 Which **way the line slopes**.

As x gets bigger, y gets **bigger**.

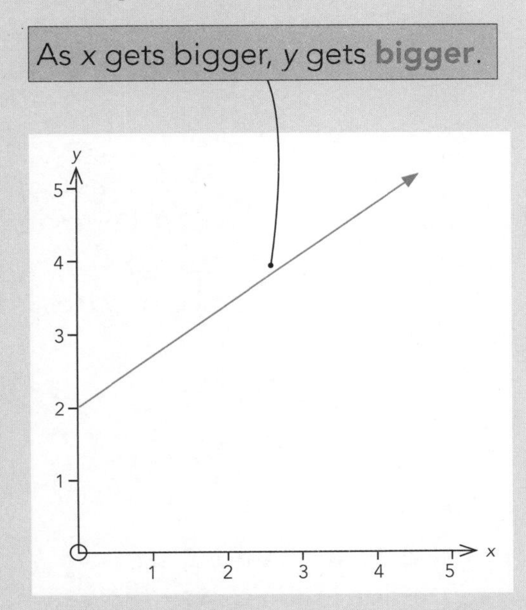

As x gets bigger, y gets **smaller**.

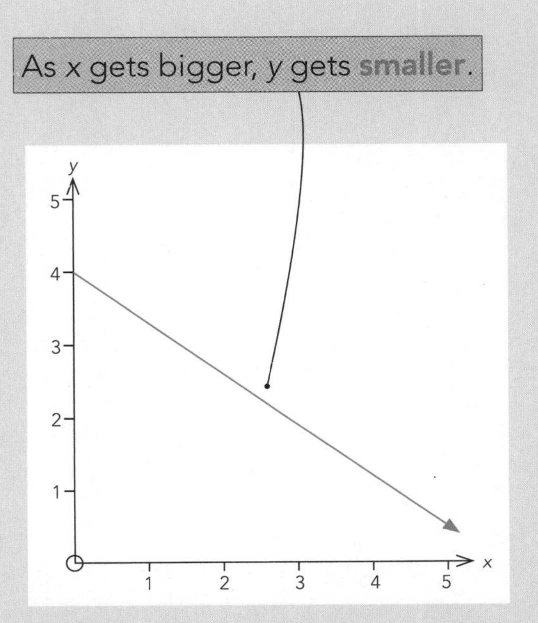

ISBN: 9780170477710

N

Examples:

1 The graph shows the relationship between mass and the cost of oranges.

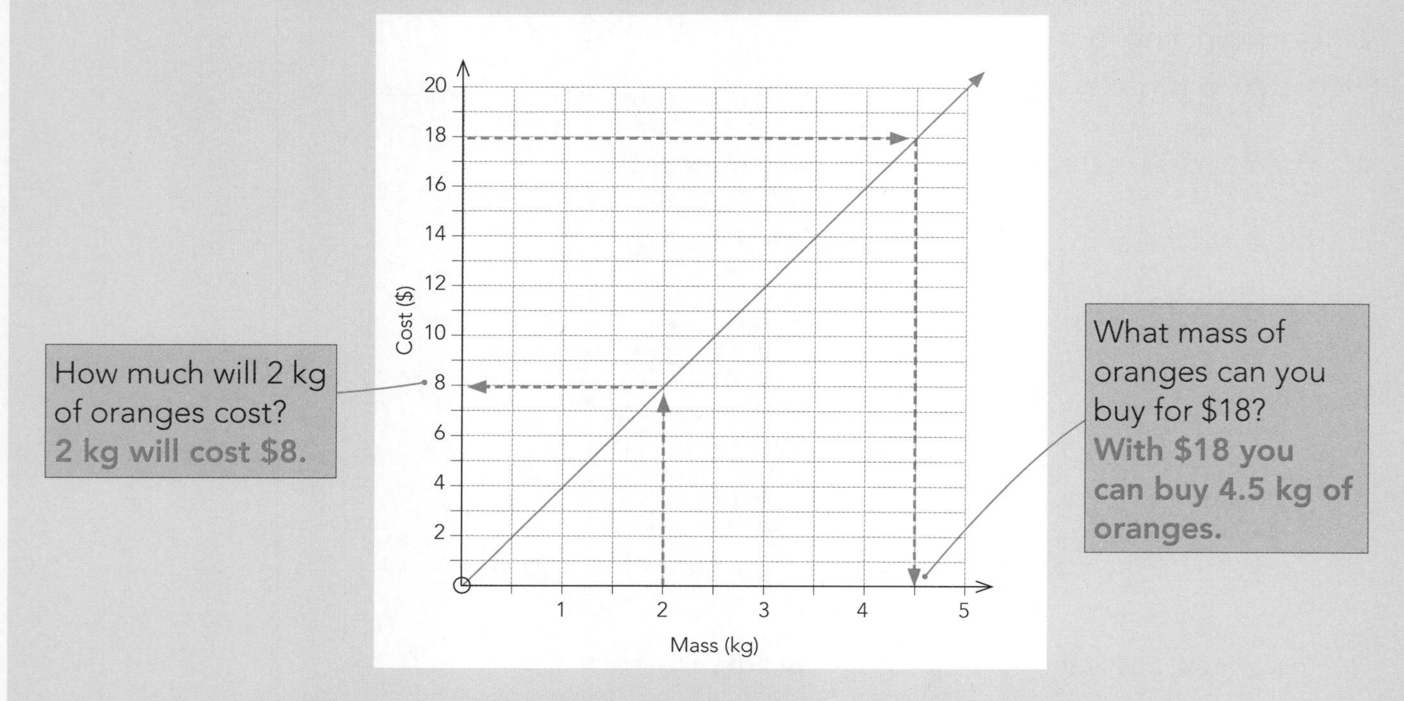

Explain how the graph would change if the price of oranges increased.
The graph still starts at the origin, but it would become steeper.

You need to write about the ***y* intercept** and the **gradient**.

2 Tatiana lives 1.2 km (1200 m) from her grandfather's home. She leaves her grandfather's and walks home at a steady speed.

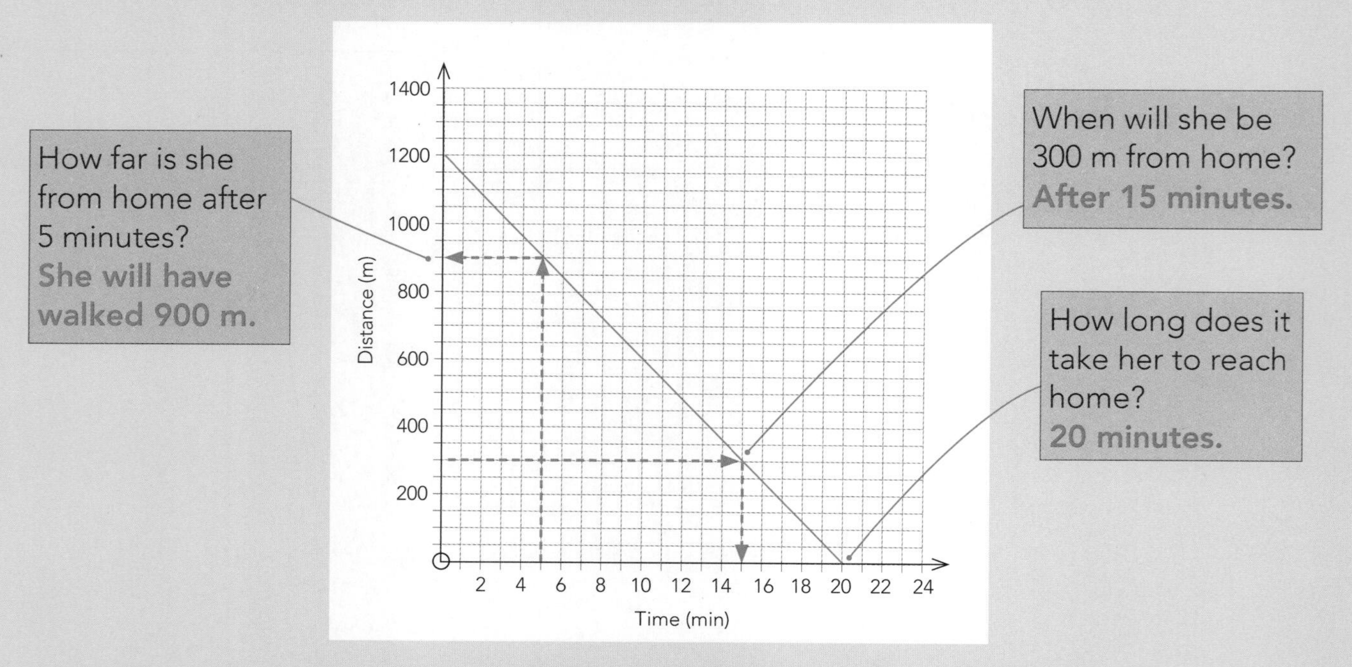

Explain how the graph would change if she lived 1.4 km from her grandfather.
The *y* intercept would increase to 1400 m, but the gradient would stay the same.

ISBN: 9780170477710

Answer these questions.

1 This graph shows the height above the ground of a hot air balloon during its ascent.

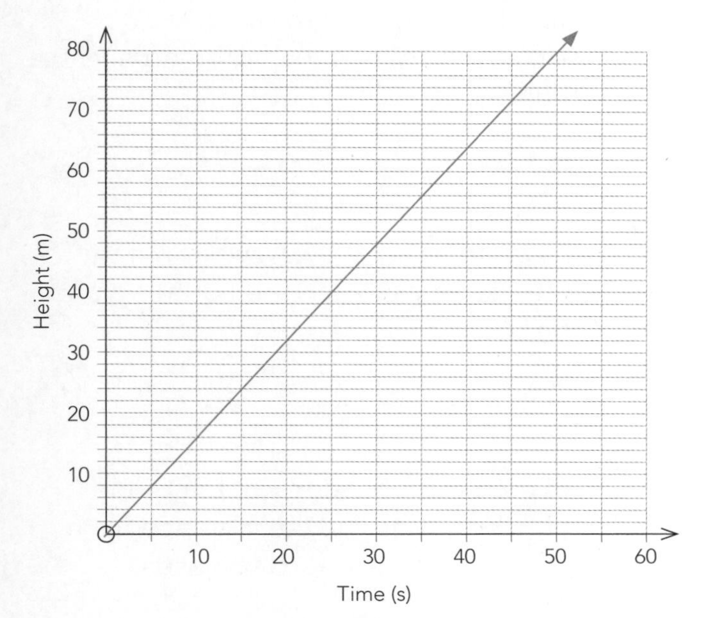

a Complete the table.

Time (s)	Height (m)
0	
10	
20	
30	
40	

How high was the balloon after

b 5 seconds? ______________ m

c 45 seconds? ______________ m

After how long is the balloon at

d 24 m? ______________ s

e 40 m? ______________ s

2 This graph shows the price of kiwifruit.

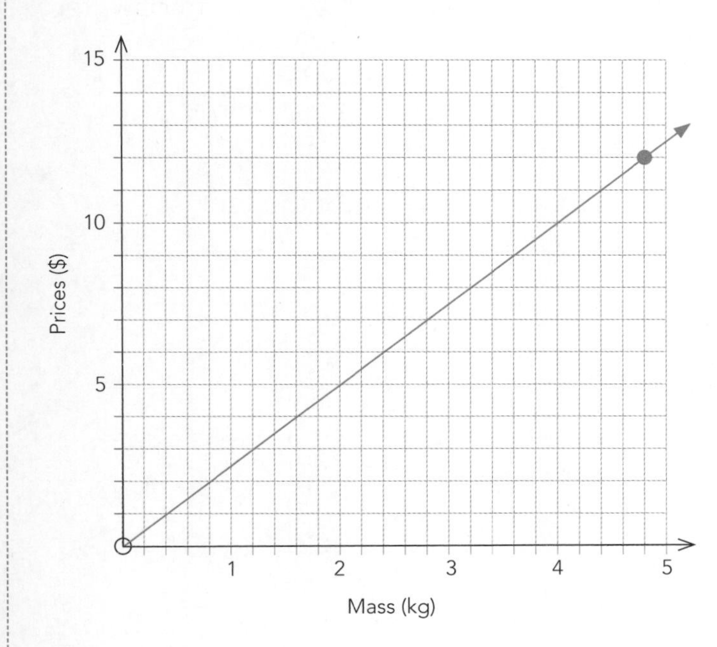

What is the cost of

a 2 kg of kiwifruit? $______________

b 3.2 kg of kiwifruit? $______________

What is the mass of a bag of kiwifruit which cost

c $10? ______________ kg

d $4? ______________ kg

e Explain what the marked point means.

____

N

3 You can work out how far away a thunderstorm is by counting the number of seconds between the lightning and the thunder. The graph shows the relationship.

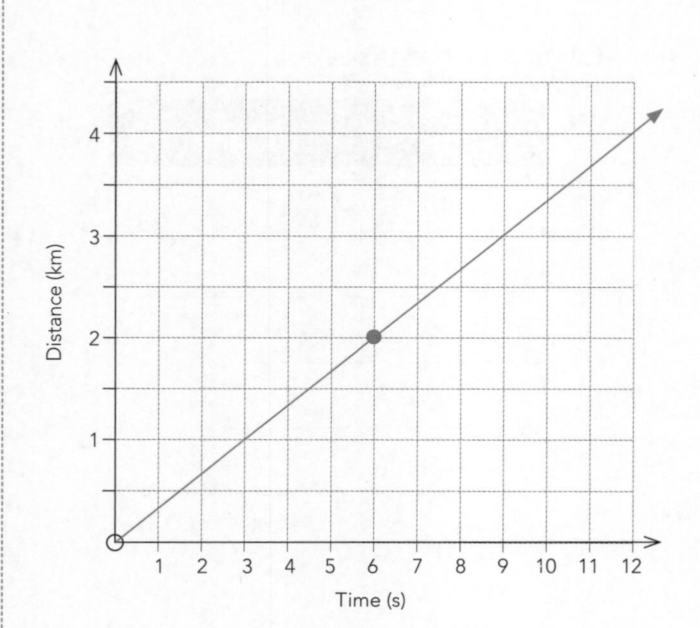

a How far away is a storm if it is 3 seconds between the lightning and the thunder?

__________________ km

b If a storm is 4 km away, what will the interval be between the lightning and the thunder?

__________________ s

c As the time between lightning and thunder increases, the distance from the storm **increases/decreases**.

d Explain what the marked point means.

____

4 The graph shows the relationship between the time a tap has been turned on and the depth of water in a tank.

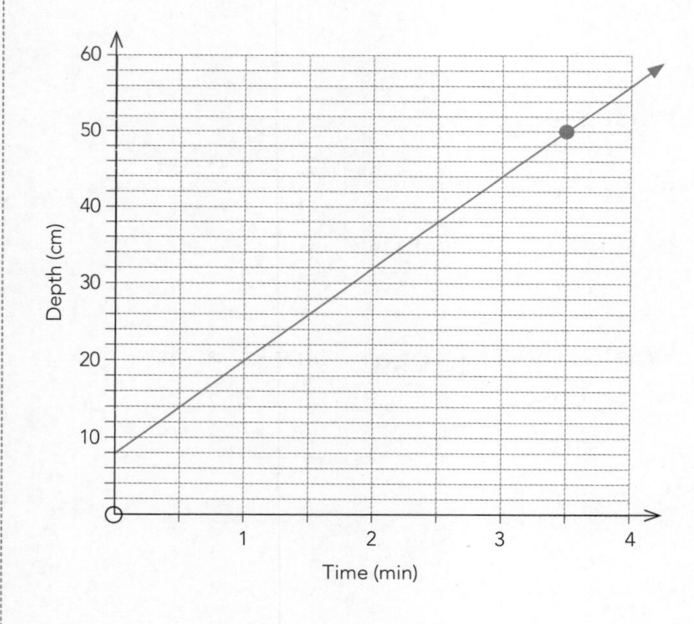

a How long will the tap have been turned on when there is 32 cm of water in the tank?

__________________ min

b What is the depth of the tank at 3 minutes?

__________________ cm

c How much water was in the tank at the start?

__________________ cm

d Explain what the marked point means.

____

ISBN: 9780170477710

5 Madi has borrowed money from her brother to buy a phone. She pays back $80 per week. The graph shows how much she owes him at the end of each week.

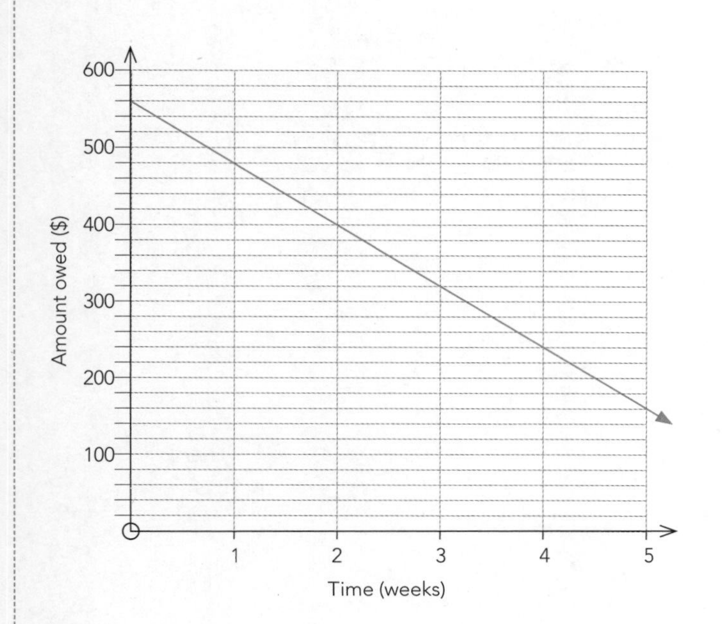

a How much did Madi borrow from her brother?

$____________________

b How much does she owe her brother after 3 weeks?

$____________________

c At the end of what week does she still owe $240 to her brother?

Week ____________________

d As the weeks go by, the amount she owes her brother **increases/decreases**.

e Explain how the graph would change if she were paying back $90 per week:

The *y* intercept would **increase/decrease/stay the same**.

The gradient would **stay the same/become steeper/become less steep**.

6 Mac is a plumber and he charges a call-out fee plus $45 for every hour he works. The graph shows how much he charges for visits of different lengths.

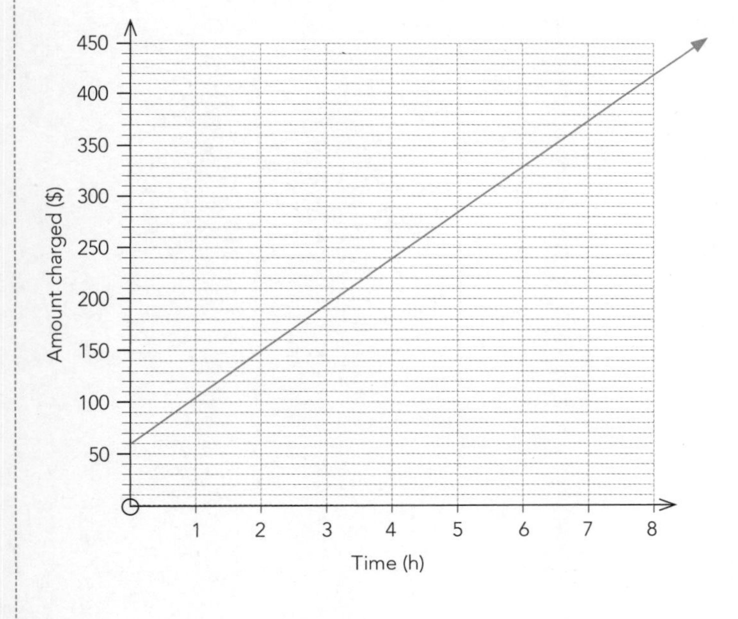

a How much is his call-out fee?

$____________________

b How much would he charge for a job that took 6 hours?

$____________________

c How long did he work if he charged $240?

____________________ hours

d As the number of hours spent on a job increases, his charge **increases/decreases**.

e Explain how the graph would change if he increased his call-out fee:

The *y* intercept would **increase/decrease/stay the same**.

The gradient would **stay the same/become steeper/become less steep**.

N

Creating graphs

Example:
It costs one New Zealand dollar to buy 1.5 Samoan tala.

Step 1: Create a table.

NZ$	Tala
40	60
100	150
200	300
300	450

Choose **at least three** points to plot.

Step 2: Plot the points on a graph.

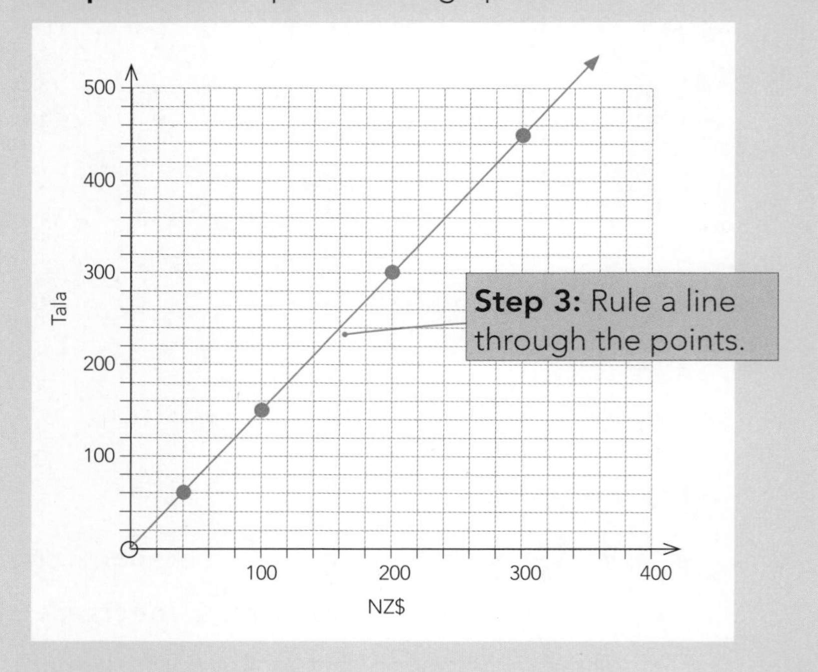

7 Apples cost $2.50 per kg.

a Complete the table.

Mass (kg)	Cost ($)
1	2.50
2	
3	
4	

b Plot the points on the graph and rule a line through them.

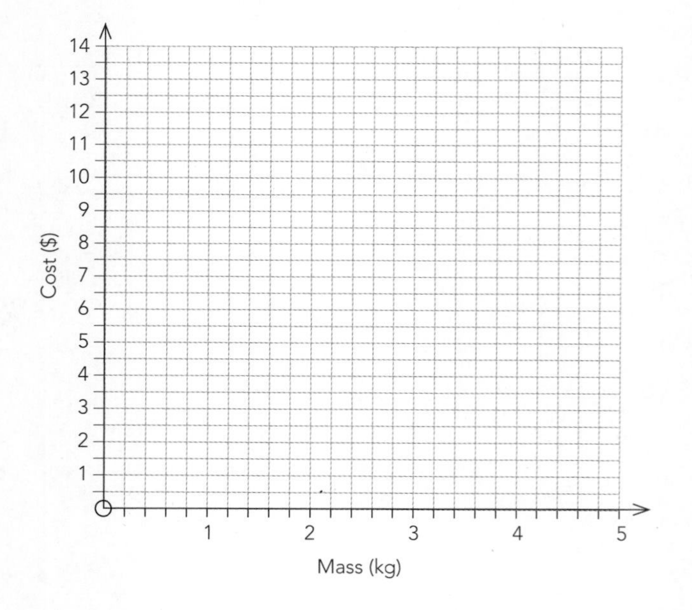

Use your graph to answer these questions.

c How much would it cost to buy 3.4 kg of apples? $__________

d How many kilograms of apples could you buy for $10.50? __________ kg

 ISBN: 9780170477710

8 Rusty Rides rents out e-bikes.
It costs $1.50 to unlock a bike and then $1 per minute.

a Complete the table.

Time (min)	Cost ($)
0	1.50
1	
2	
3	
4	

b Plot the points on the graph and rule a line through them.

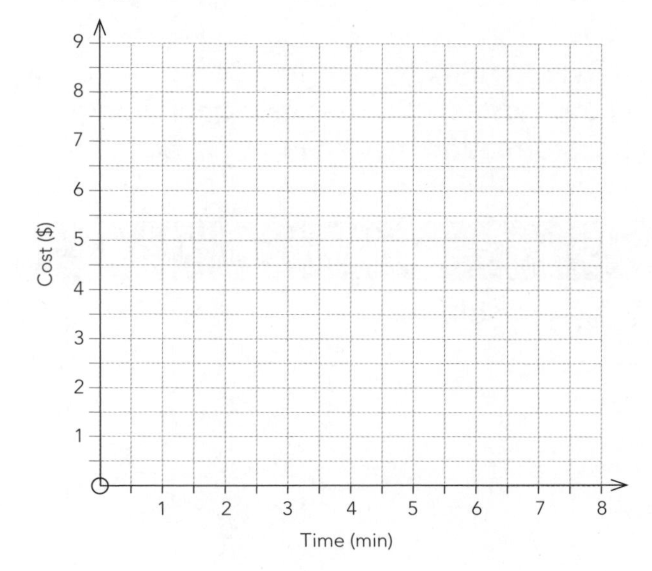

Use your graph to answer these questions.

c How much will it cost to hire a bike for seven and a half minutes? $__________

d For how long could you hire a bike for $7? __________ min

9 Leah lives 2200 m from her school. She walks at a constant rate of 75 m per minute to get to school.

a Complete the table.

Time (min)	Distance from school (m)
0	2200
4	
8	
12	

b Plot the points on the graph and rule a line through them.

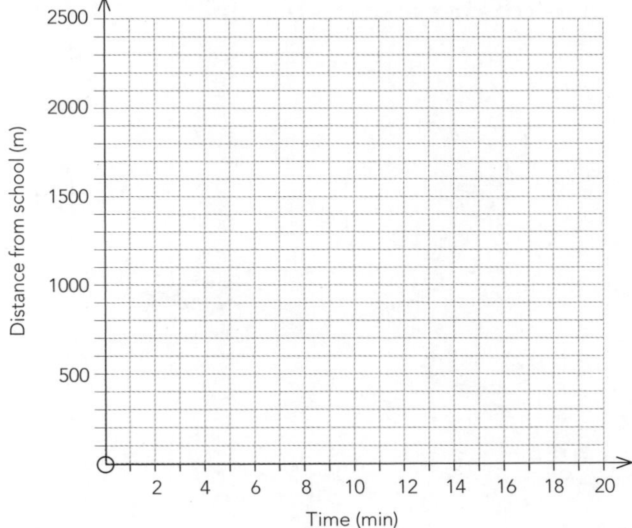

Use your graph to answer these questions.

c How far will she be from school after she has walked for 20 minutes? __________ m

d When will she be 850 m from school? __________ min

Spatial properties and representations

Transformation geometry

- A transformation is changing or moving a figure by following certain rules.
- The transformed figure is called the **image**.
- A transformation may change the position, size, shape or orientation of the object.

Translation	Reflection
The figure is **shifted**. 	The figure is **reflected in a mirror** line. 
Rotation	**Enlargement**
The figure **rotates around a point**. 	The figure **gets bigger or smaller**. 

ISBN: 9780170477710

Highlight the transformations used in the following pairs of images.

1

Translated Reflected
Rotated Enlarged

2

Translated Reflected
Rotated Enlarged

3

Translated Reflected
Rotated Enlarged

4

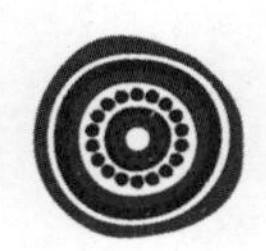

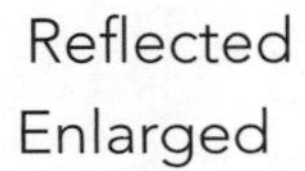

Translated Reflected
Rotated Enlarged

5

Translated Reflected
Rotated Enlarged

6

Translated Reflected
Rotated Enlarged

7

Translated Reflected
Rotated Enlarged

8

Translated Reflected
Rotated Enlarged

ISBN: 9780170477710

N

Reflection

- The figure is **reflected in a mirror** line.

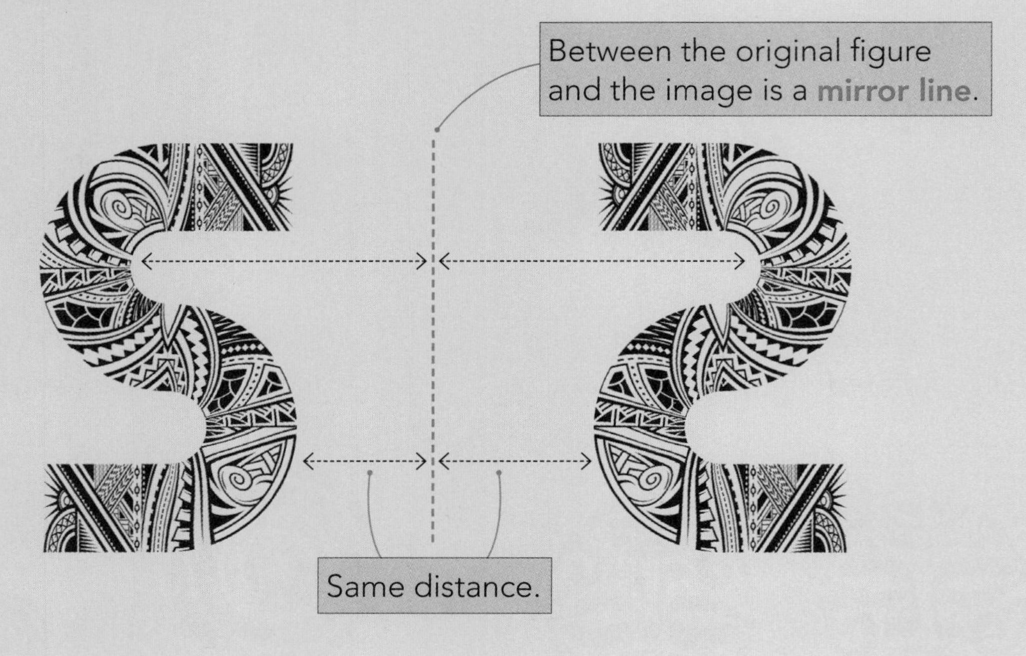

Draw mirror lines for the following reflections.

9

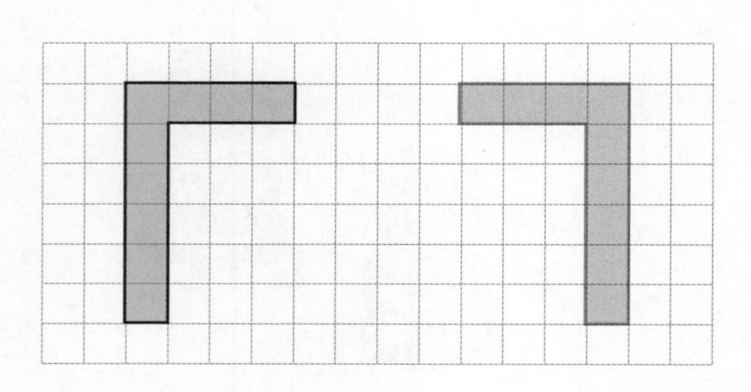

10

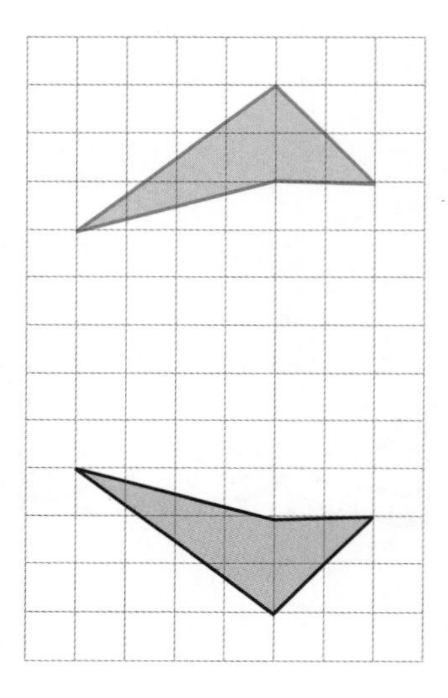

11

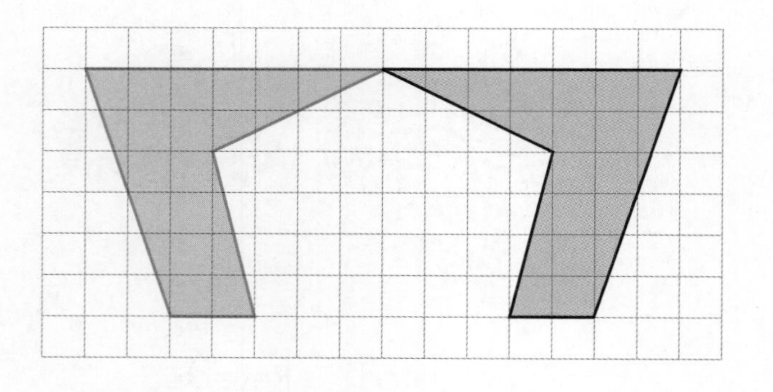

12

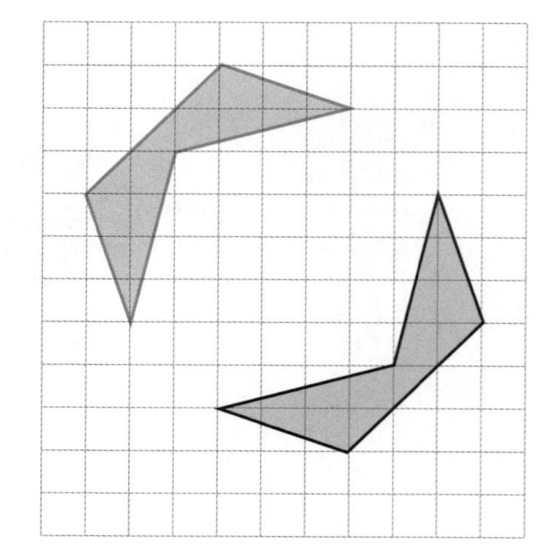

ISBN: 9780170477710

Line symmetry

- Some figures form reflections of themselves in a mirror line.
- They are said to have **line symmetry**.
- **Each mirror line** that can be drawn through a figure is known as **a line of symmetry**.
- The **order** of line symmetry is the **total number of lines of symmetry** in a figure.

This figure has **one** line of symmetry.

Order of line symmetry = 1

This figure has **six** lines of symmetry.

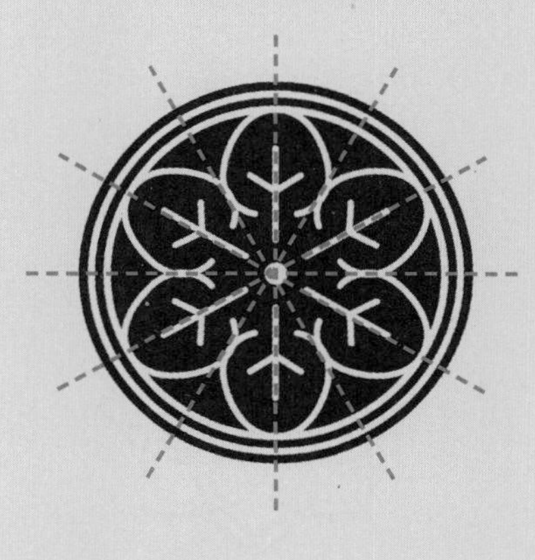

Order of line symmetry = 6

Draw all the lines of symmetry on these figures and write the order of line symmetry for each.

13

Order of line symmetry = ________

14

Order of line symmetry = ________

15

Order of line symmetry = ________

16

Order of line symmetry = ________

ISBN: 9780170477710

N

17

Order of line symmetry = ________

18

Order of line symmetry = ________

19

Order of line symmetry = ________

20

Order of line symmetry = ________

21

Order of line symmetry = ________

22

Order of line symmetry = ________

23

Order of line symmetry = ________

24

Order of line symmetry = ________

 ISBN: 9780170477710

Rotation

- The figure **rotates around a point**.
- The **point** is known as the **centre of rotation**.
- Rotations, unless specified otherwise, are always measured in a **clockwise direction**.

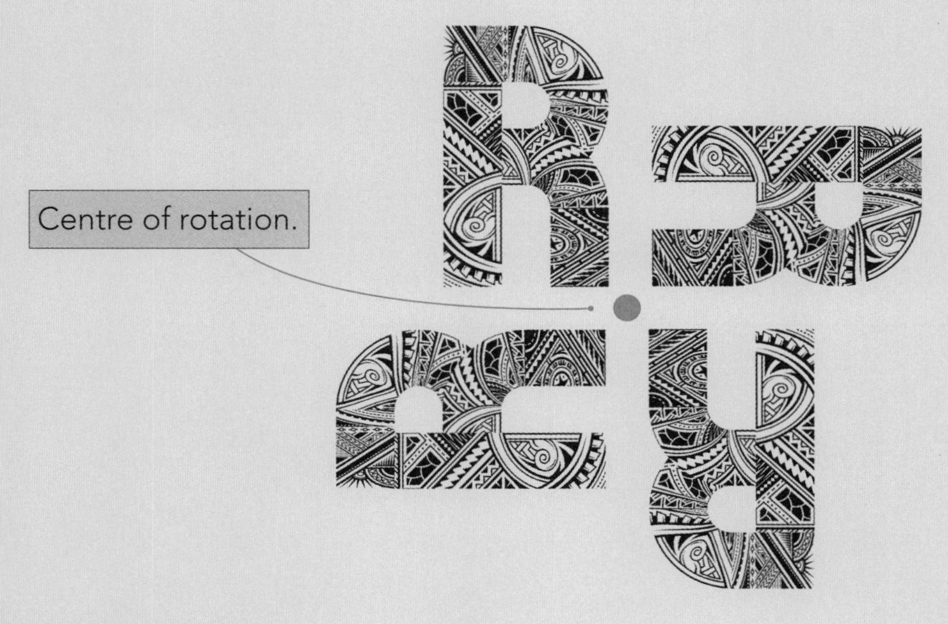

Angles of rotation

- A **full rotation** is a rotation through **360°**.

1 Turns using right angles

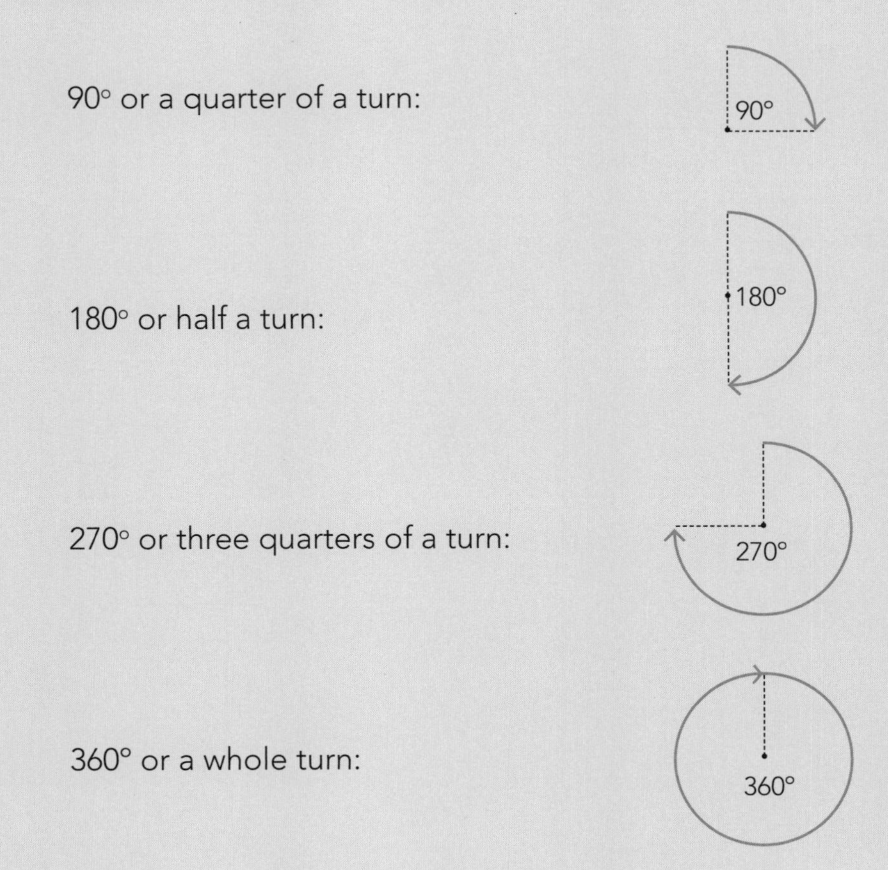

N

Using angles of rotation

Examples: Write the angle of rotation for each of the following.

Remember: 1 The **green** figure is the **original**, so **start** your angle there.
2 The **grey** figure is the **image**.

Remember: Unless you are told otherwise, angles are always measured in a **clockwise** direction.

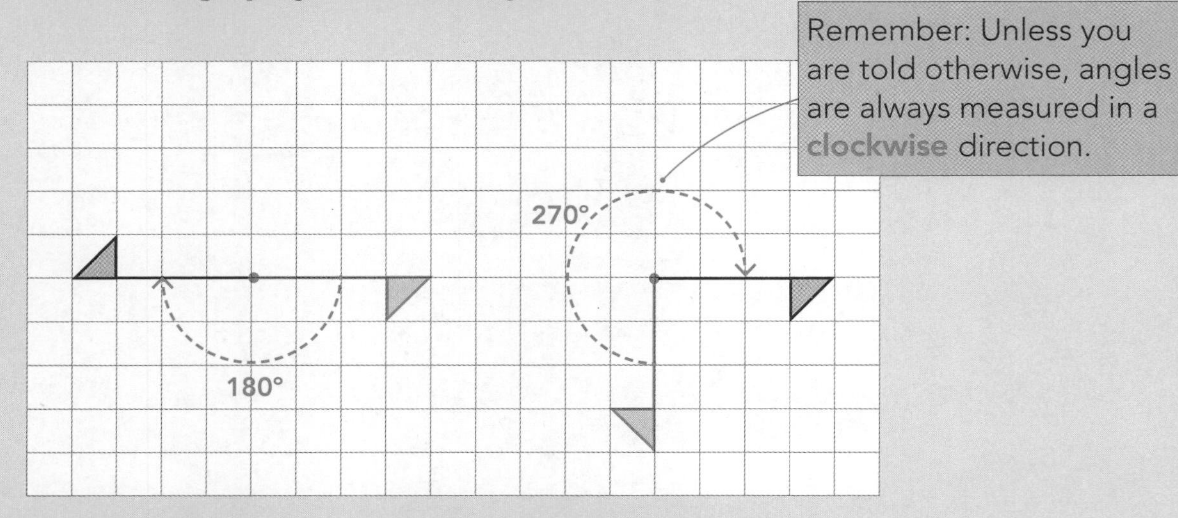

Write the angle of rotation for each of the following.

25

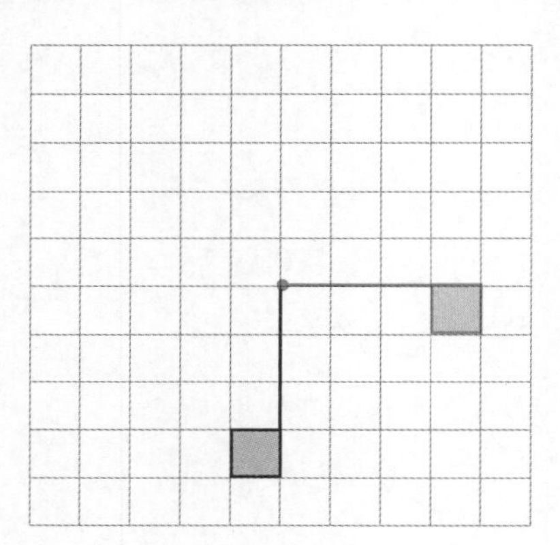

Angle = ________°

26

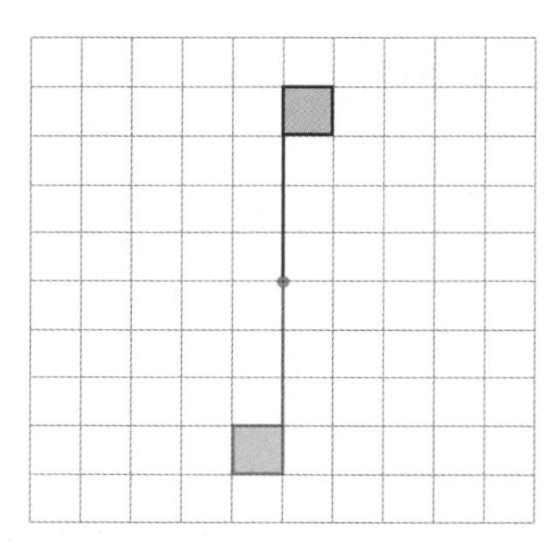

Angle = ________°

27

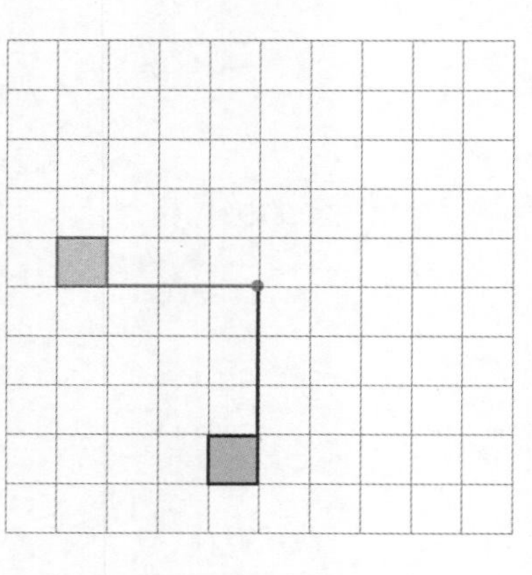

Angle = ________°

28

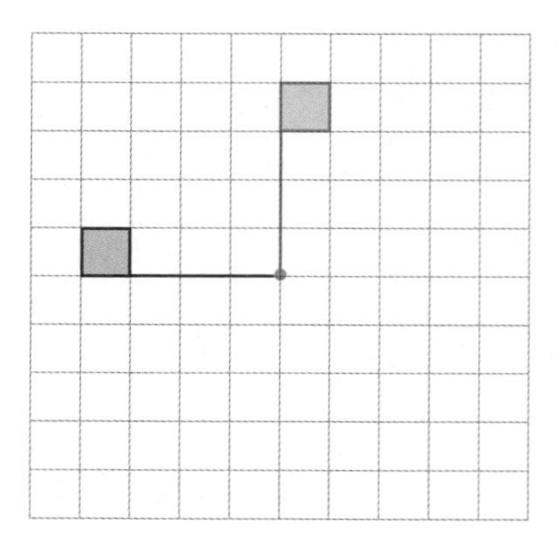

Angle = ________°

ISBN: 9780170477710

2 Turns that do not use right angles

Examples: Calculate the angles between the arrows in these patterns.

1

$$\text{Angle} = \frac{360°}{3} = 120°$$

There are 3 angles in one rotation.

2

$$\text{Angle} = \frac{360°}{6} = 60°$$

There are 6 angles in one rotation.

Calculate the angle of rotation for the leaves or petals in each of the following.

29

Angle = $\frac{360°}{\square}$

= ________ °

30

Angle =

= ________ °

31

Angle =

= ________ °

32

Angle =

= ________ °

Rotational symmetry

- All figures can be rotated so they map onto themselves.
- All figures have at least one order of **rotational symmetry**; some have more than one.
- The **order of rotational symmetry** is the **number of times a figure looks the same** (maps onto itself) during a full rotation (360°).

Examples:

1 This figure looks the same **once** during a full rotation.

This figure looks the same as the original.

This means that its order of rotational symmetry is **1**.

2 This figure looks the same **twice** during a full rotation.

These two figures look the same as the original.

This means that its order of rotational symmetry is **2**.

State the order of rotational symmetry for these images.

33

Order of rotational symmetry = __________

34

Order of rotational symmetry = __________

ISBN: 9780170477710

35

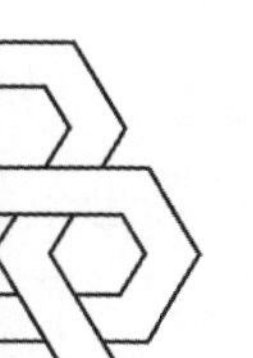

Order of rotational symmetry = __________

36

Order of rotational symmetry = __________

37

Order of rotational symmetry = __________

38

Order of rotational symmetry = __________

39

Order of rotational symmetry = __________

40

Order of rotational symmetry = __________

41

Order of rotational symmetry = __________

42

Order of rotational symmetry = __________

ISBN: 9780170477710

N

Enlargement

- The figure **gets bigger or smaller**.
- The **scale factor** tells us how much bigger or smaller the figure becomes.

$$\text{scale factor} = \frac{\text{length of image}}{\text{length of original figure}}$$

Examples:

The lines in the image are **three times** as long as those in the original. The image has a scale factor of **3**.

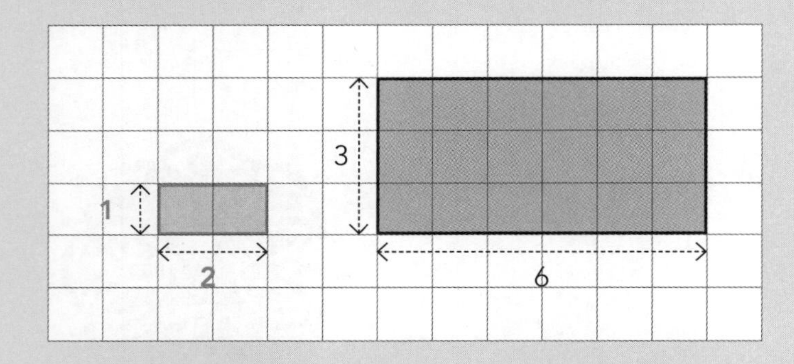

$$\text{scale factor} = \frac{3}{1} = \frac{6}{2} = 3$$

The lines in the image are **half** as long as those in the original. The image has a scale factor of **half** or **0.5**.

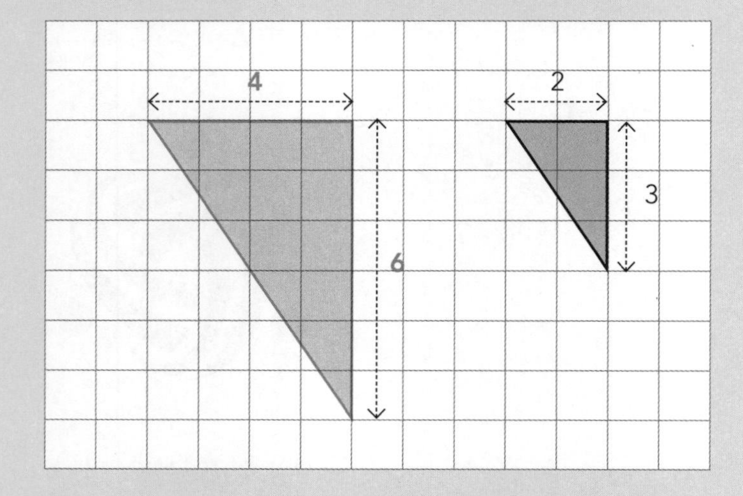

$$\text{scale factor} = \frac{2}{4} = \frac{3}{6} = \frac{1}{2}$$

ISBN: 9780170477710

Write the scale factor for these enlargements.

43

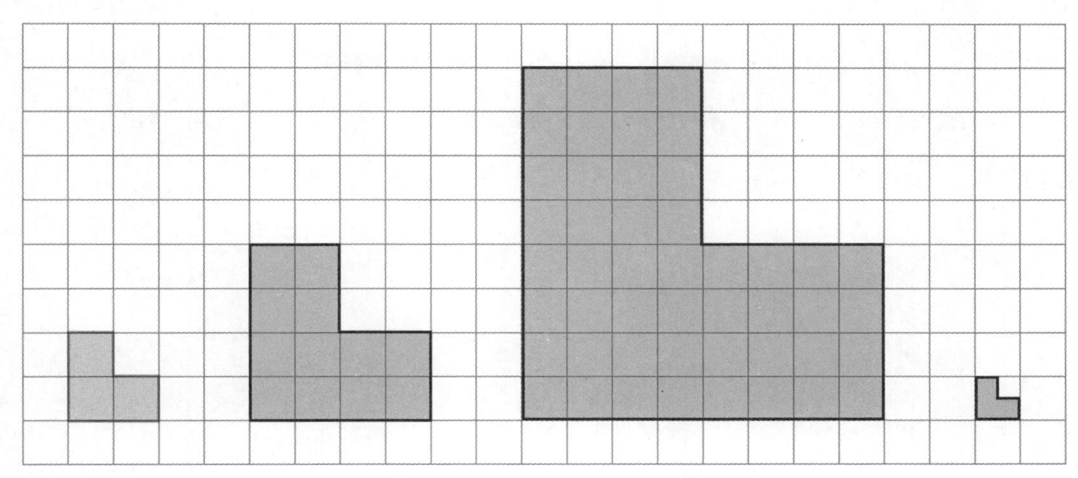

scale factor $= \frac{4}{2} = \frac{2}{1} =$ ______

scale factor = ―― = ―― = ______

scale factor = ______

44

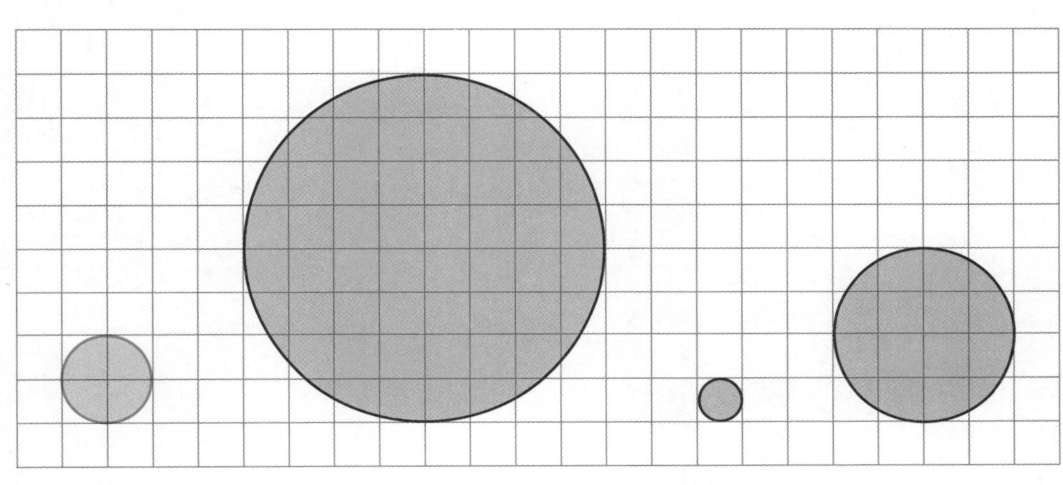

scale factor = ______

scale factor = ______

scale factor = ______

45

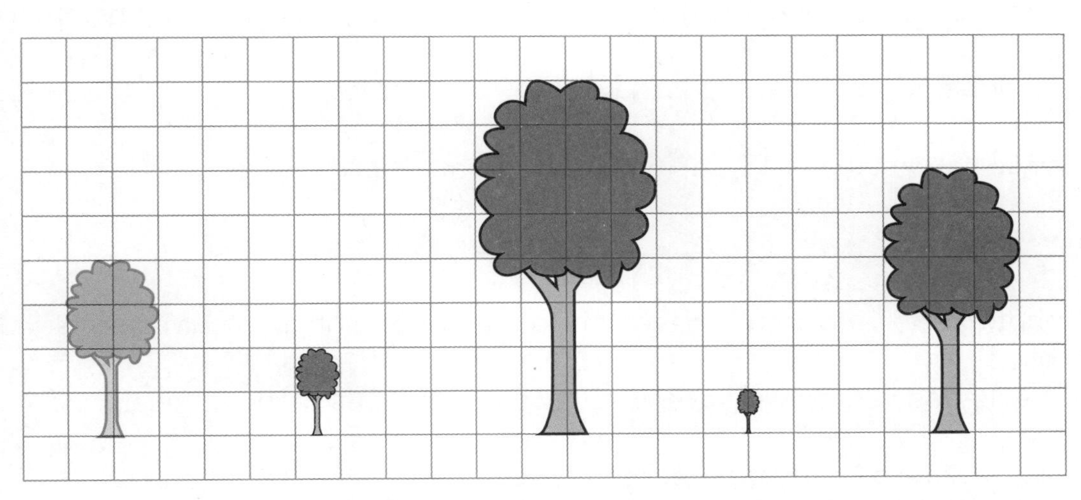

scale factor = ______

scale factor = ______

scale factor = ______

scale factor = ______

ISBN: 9780170477710

Areas after enlargement

- So far we have considered the scale factors for the **lengths** of figures.

The area of an image = the area of the original x (scale factor)^2

Examples:

1

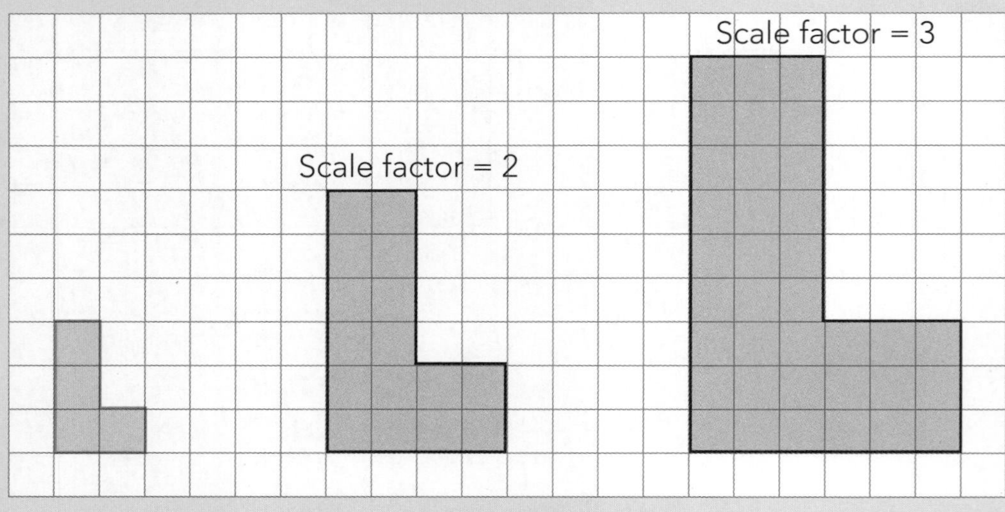

Area = 4 squares

Area = **4** x scale factor^2
= **4** x 2^2
= 16 squares

Area = **4** x scale factor^2
= **4** x 3^2
= 36 squares

Check by counting the squares.

2 A circle has an area of 25 cm^2. Its radius is tripled (multiplied by 3).

Its new area = 25 x scale factor^2
= 25 x 3^2
= 225 cm^2

Answer the following questions.

46 A rectangle has an area of 20 cm^2. Calculate its area if its length and height are

a doubled: ____________ cm^2 **b** tripled: ____________ cm^2

47 A circle has an area of 15 cm^2. Calculate its area if its radius is

a doubled: ____________ cm^2 **b** multiplied by 5: ____________ cm^2

48 A photograph is enlarged so that it is four times as high and four times as wide as the original.
The enlarged version will have ________ times the area of the original.

49 Ana is making a courtyard. Her brother tells her that it will not be big enough, and she should double the length and the width.
If the original design needed 60 tiles, how many tiles would she need? ____________ tiles

ISBN: 9780170477710

2D and 3D representations

Dimensions

1D (one-dimensional) shapes	2D (two-dimensional) shapes	3D (three-dimensional) shapes
• have 1 measurement: length. 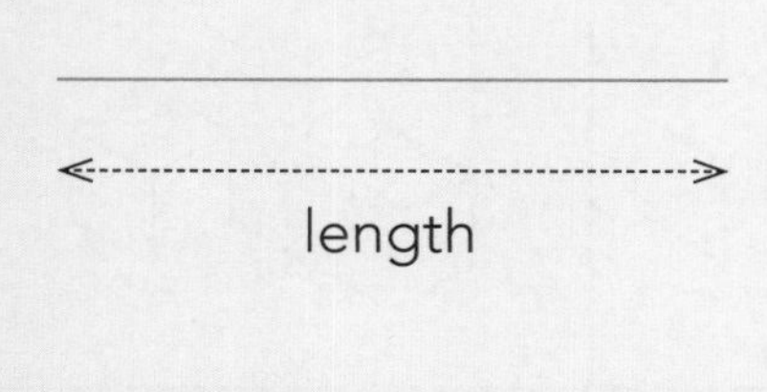	• have 2 measurements: height and width. 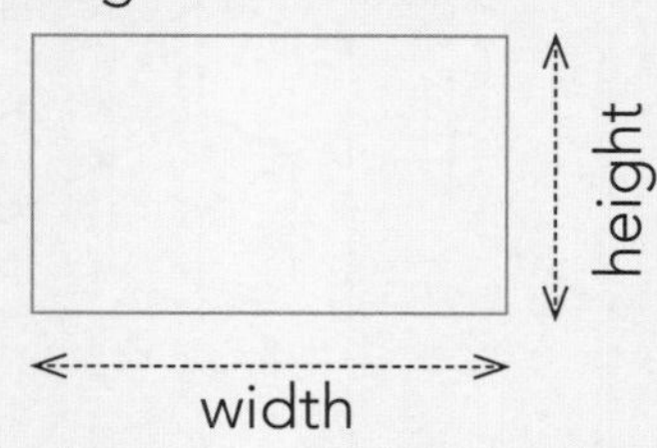	• have 3 measurements: height, width and depth. 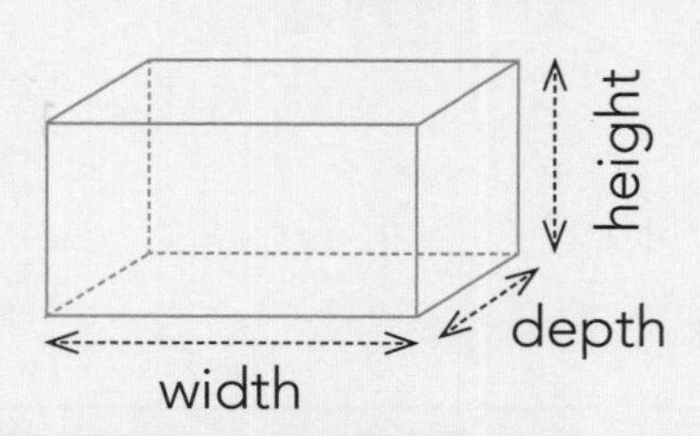

Vertices, faces and edges (sides)

- We talk about **one vertex** or **several vertices**.

2D shapes have vertices and sides.

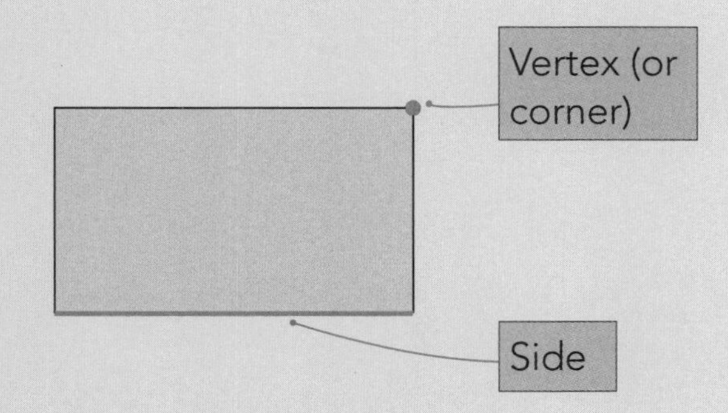

A rectangle has **4 sides** and **4 vertices**.

3D shapes have vertices, faces and edges. The faces are 2D shapes.

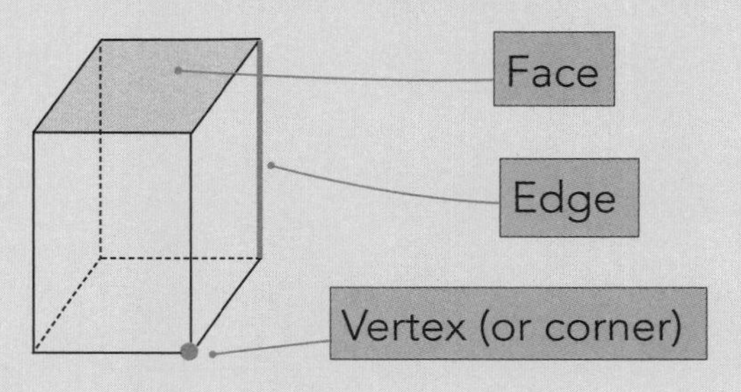

A cuboid has **8 faces**, **8 vertices** and **12 edges**.

Different views of 3D shapes

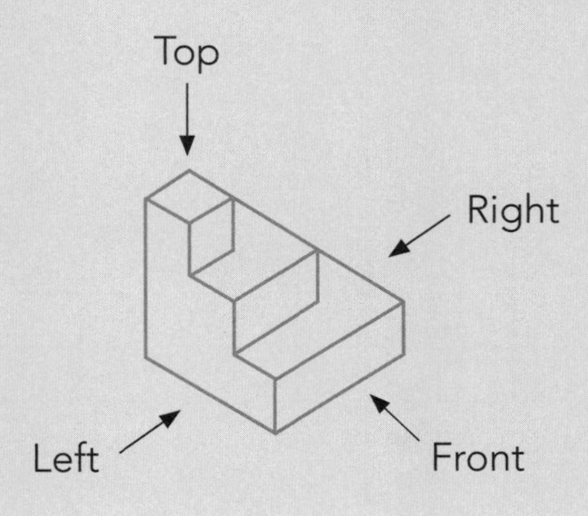

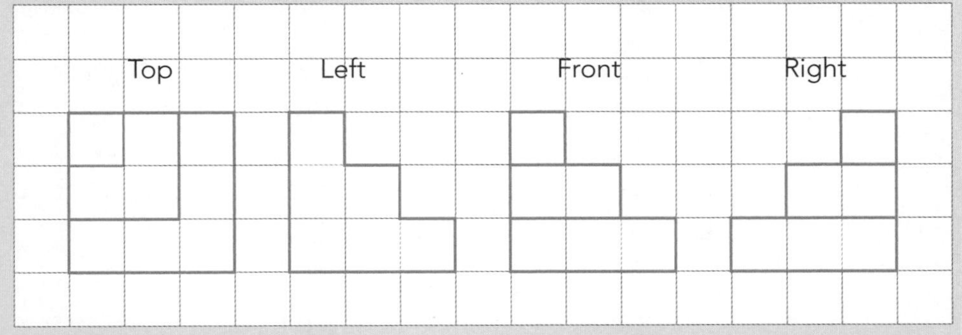

ISBN: 9780170477710

N

Different views of 3D shapes

Match these shapes with their 2D views. There are no blocks hidden in behind.

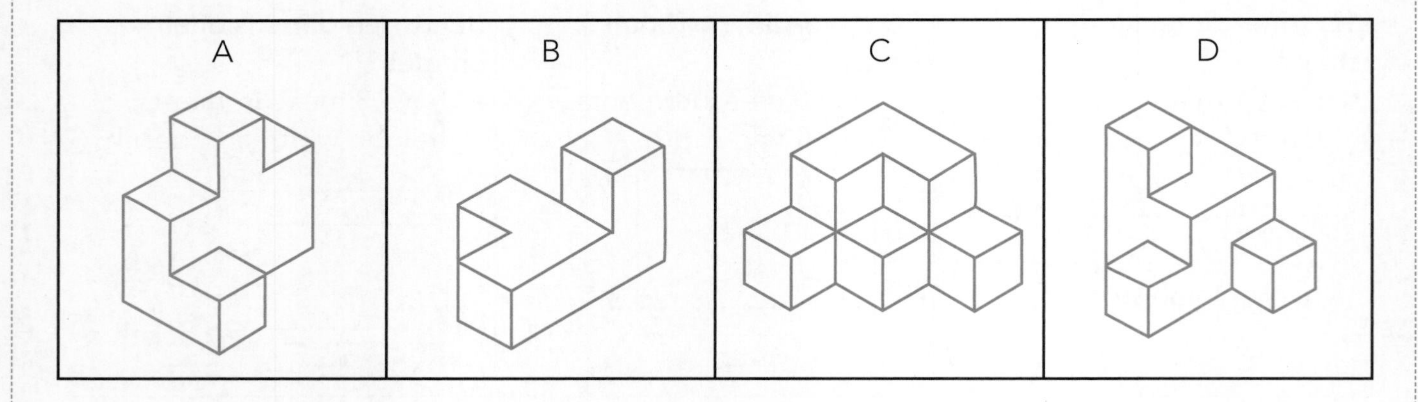

1

______

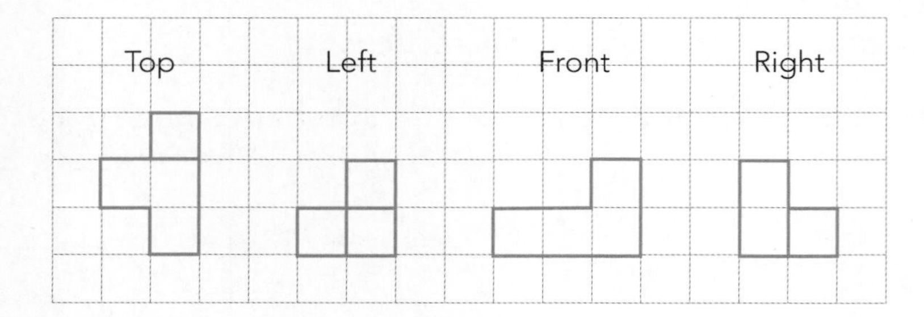

2

______

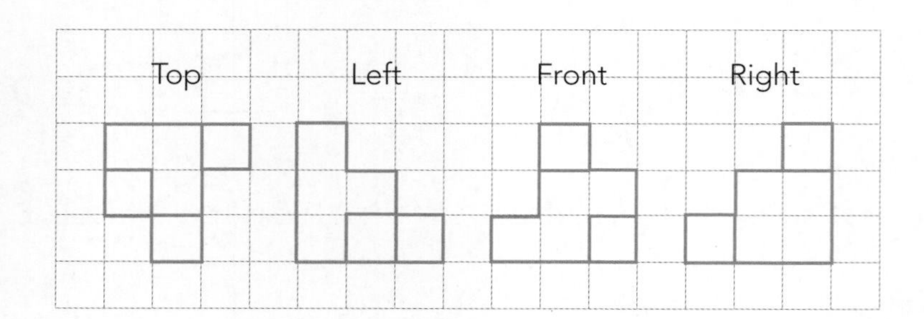

3

______

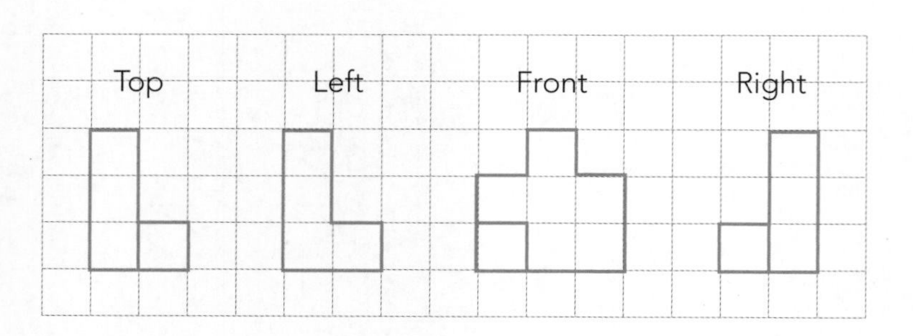

4

______

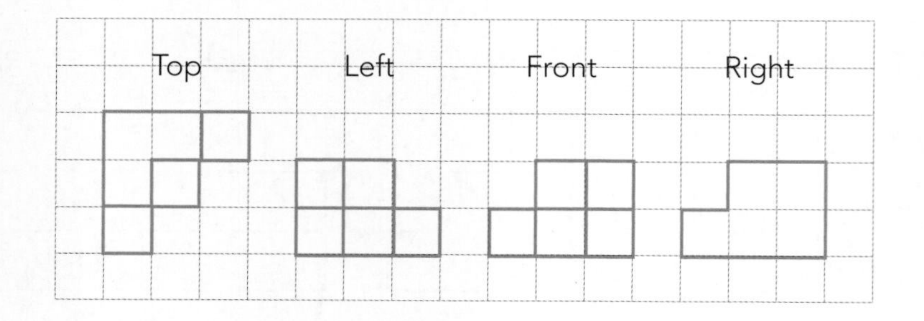

ISBN: 9780170477710

Answer the following questions.

5 Henri built this shape using cubes.

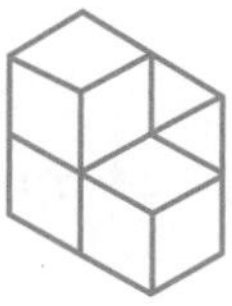

a How many more cubes would he need to add in order to make this shape?

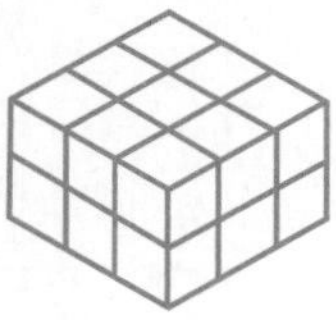

______ cubes

b If his first shape was sat on some newspaper and sprayed in pink paint, how many faces of the blocks would be pink?

______ faces

6 The only window in Tai's playhouse is beside the door. He is standing on the far side of the house. Which image of the house shows what he would see?

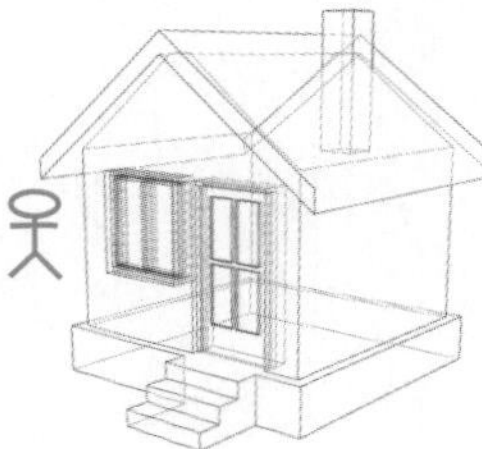

A 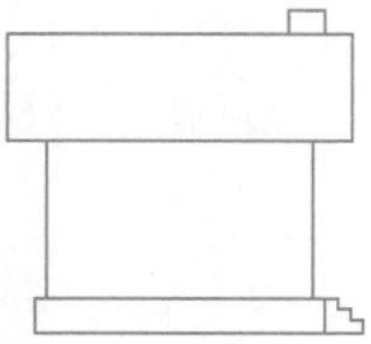 B 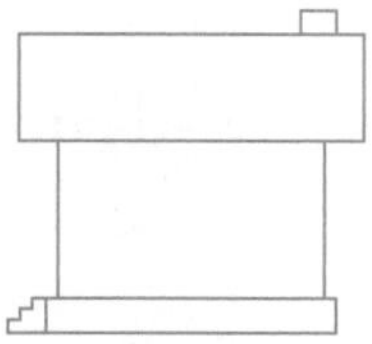 C 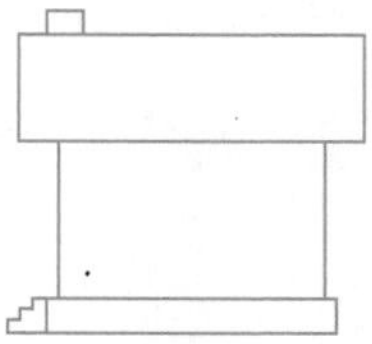 D

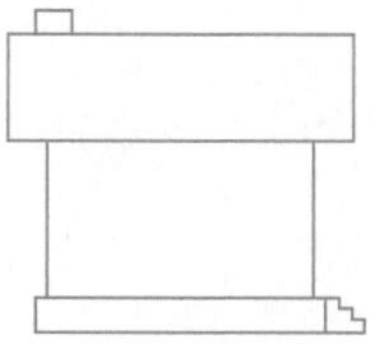

7 Jan is looking for kitchen chairs. She stands in front of this chair and walks around it in a clockwise direction. List the views of the chair in the order that she would see them during a complete circuit.

A

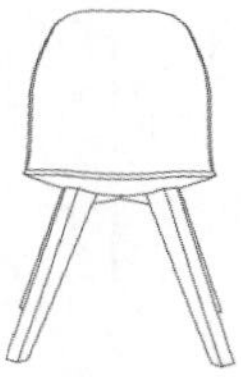

A							

B C D E F G H

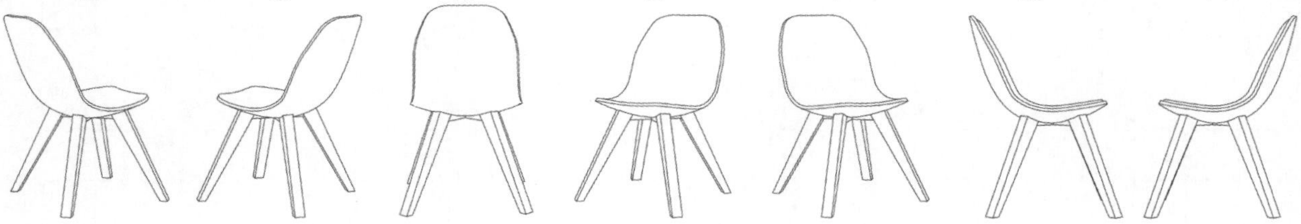

8 Adam is looking at a die (plural 'dice') which is sitting on a glass table. If he looked at the die from the opposite side and up through the table, which view would he see?
Hint: The numbers on opposite faces of a die add to 7.

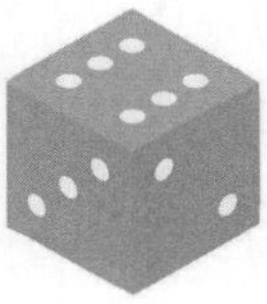

View ____________

A B 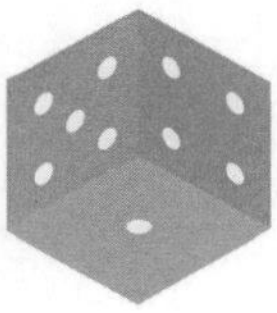 C 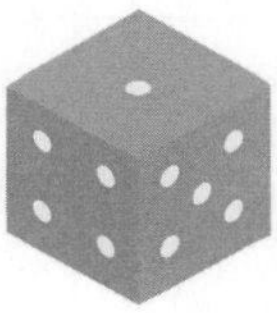 D

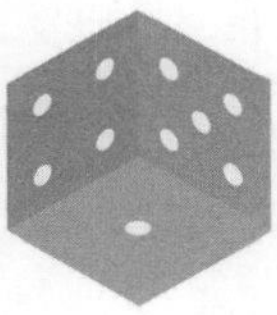

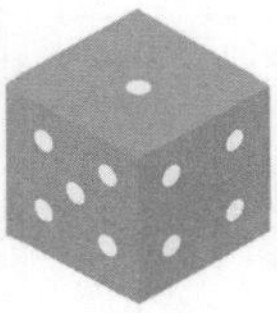

N

Nets

- Nets are the two-dimensional version of three-dimensional shapes. When cut out and folded correctly, they form a solid.

Example: This solid is a cube and can be constructed from a net like this:

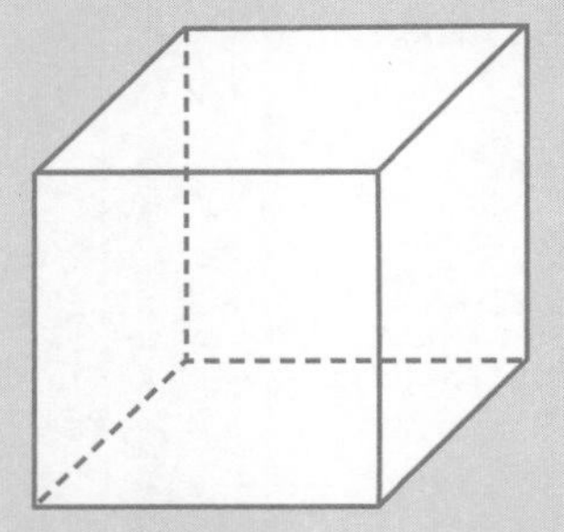

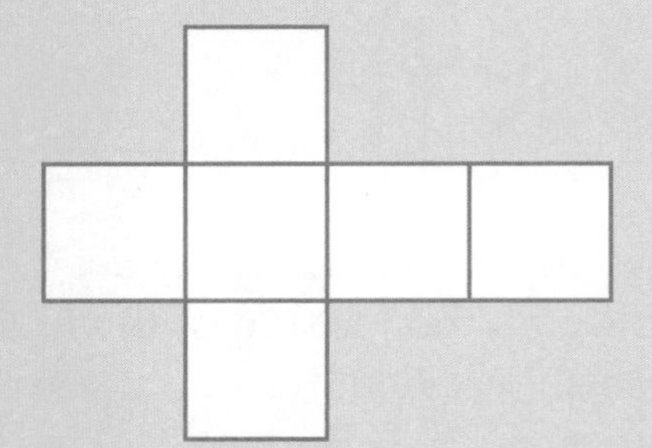

1 Highlight or shade the nets that could make a closed cube. You should find 11.

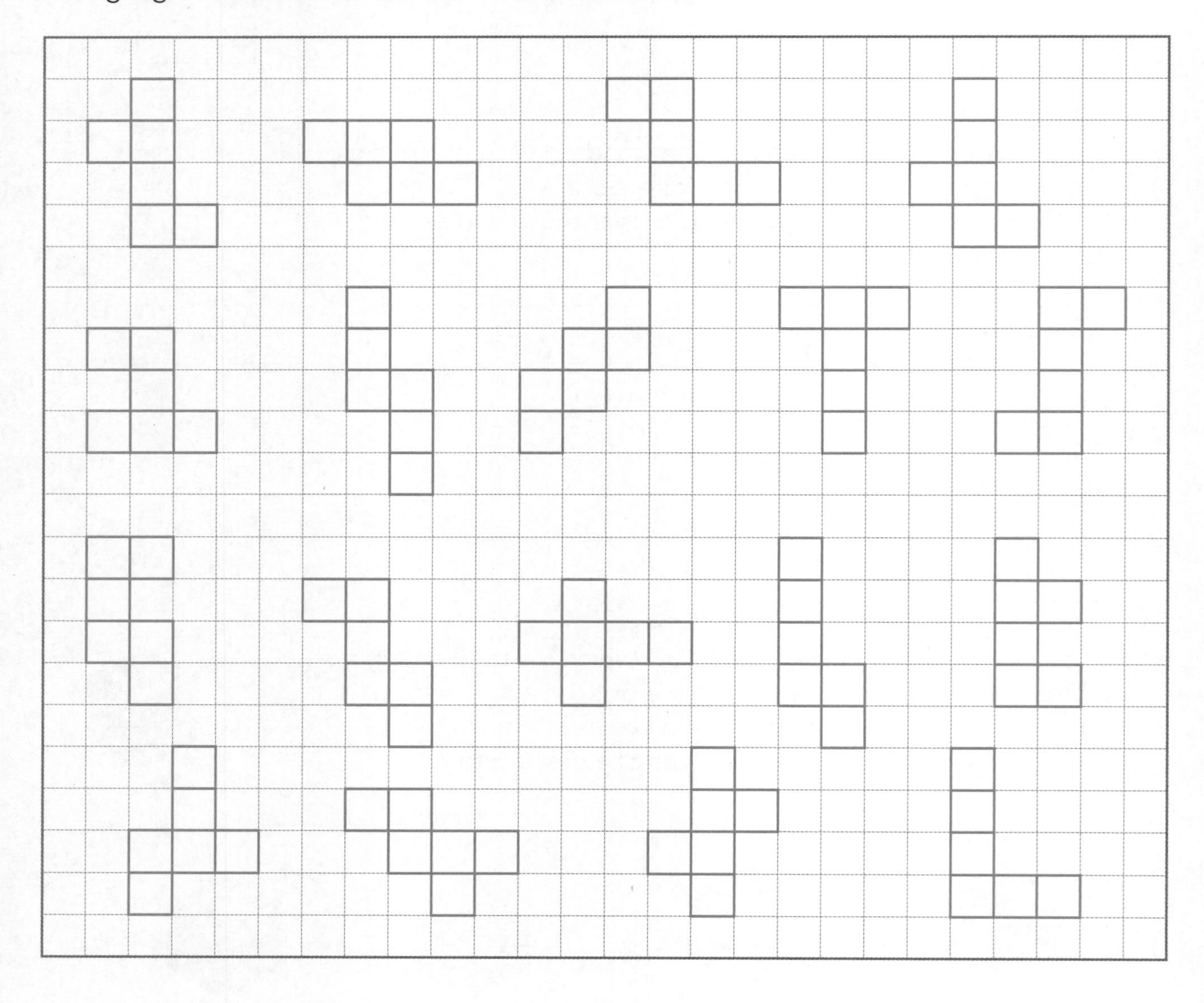

ISBN: 9780170477710

2 All of these nets can fold to form cubes.

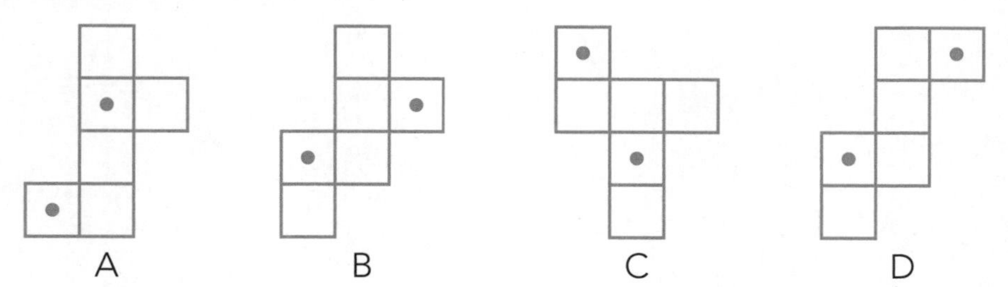

Which cube would **not** have dots on opposite faces? ____________

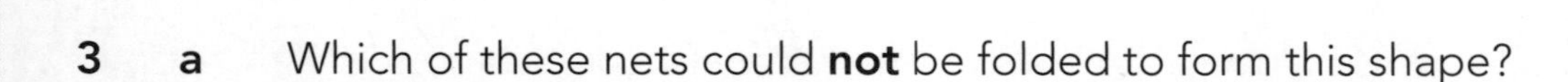

3 **a** Which of these nets could **not** be folded to form this shape?

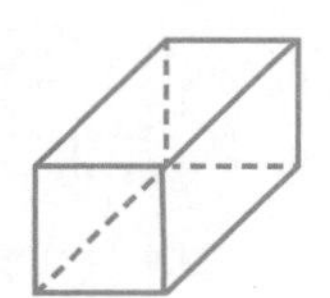

____

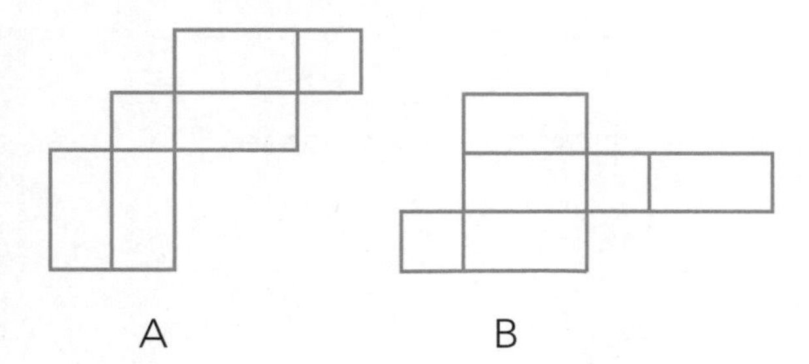

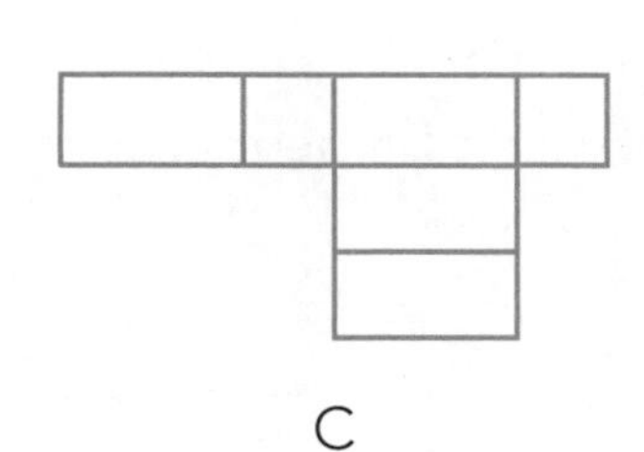

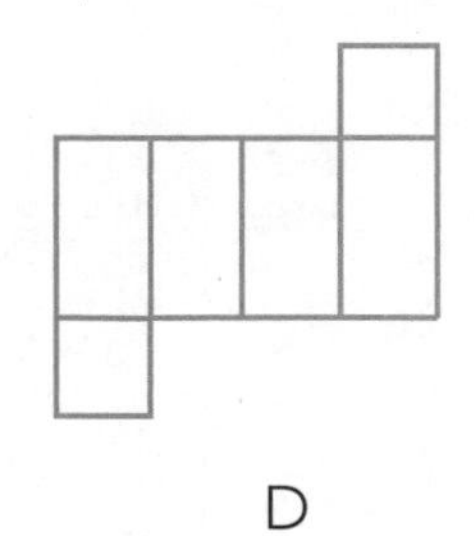

Highlight the edges that will connect to the dots when the box is made.

b

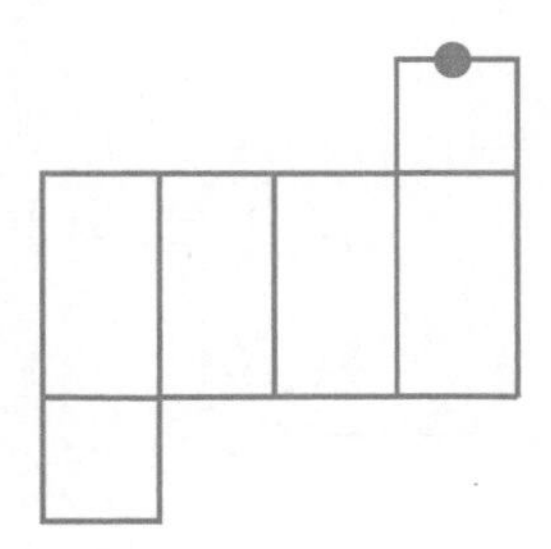

c

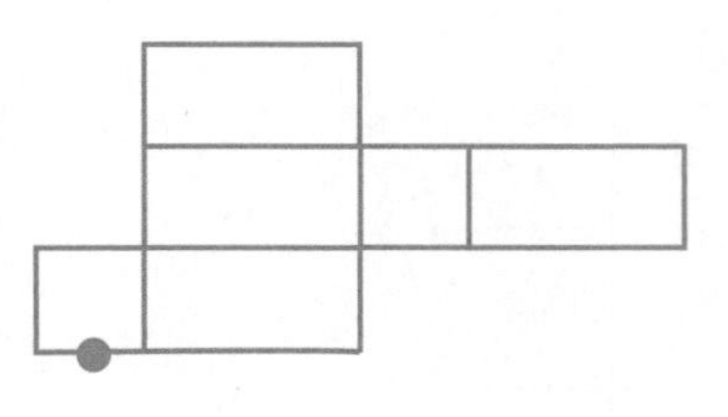

4 Kirsty is making this shape:

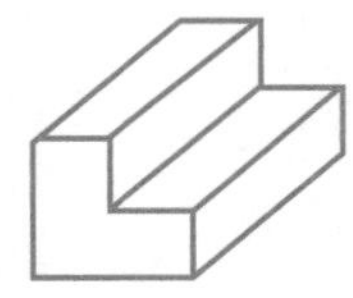

Highlight the edges that will connect to the dots when the shape is made.

a

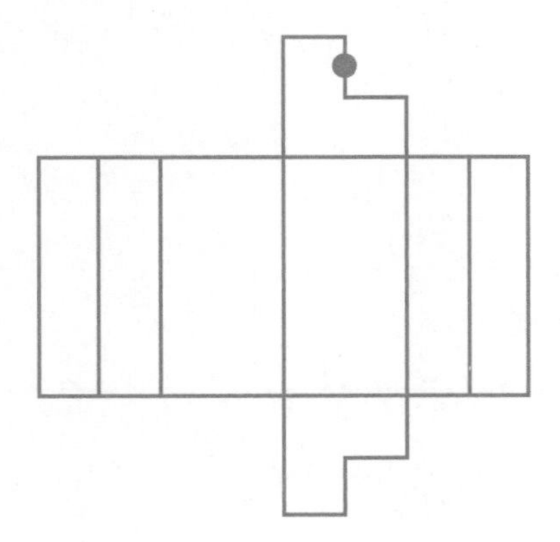

b

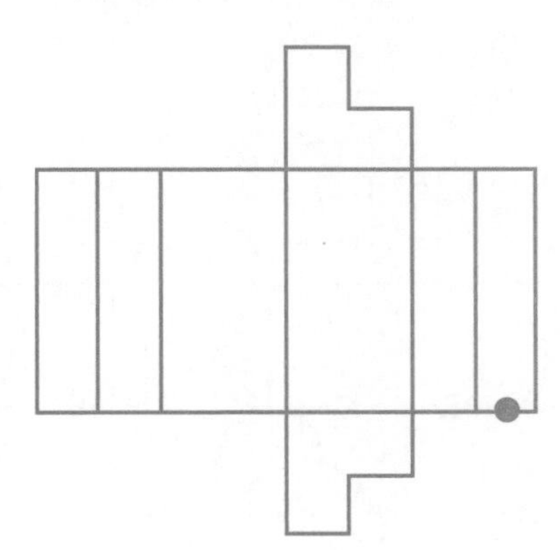

N

5 Tick the boxes for the nets which could be folded to form this shape.

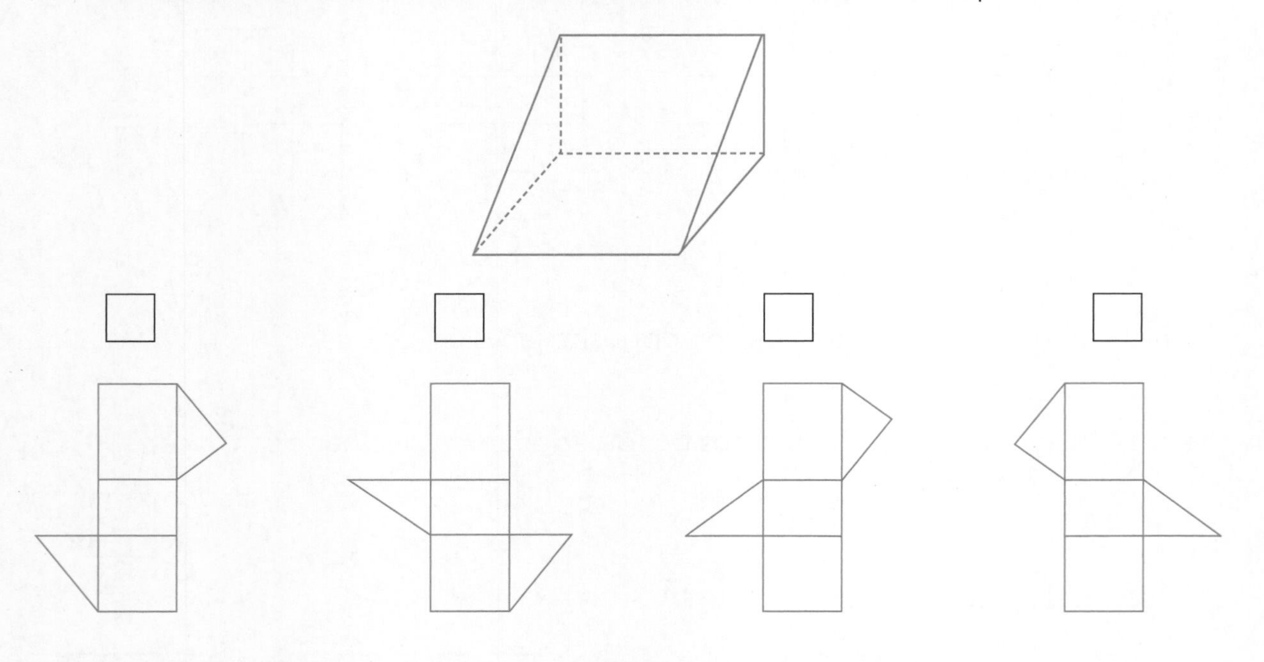

6 One of these cubes could be made from this net. Tick the correct cube.

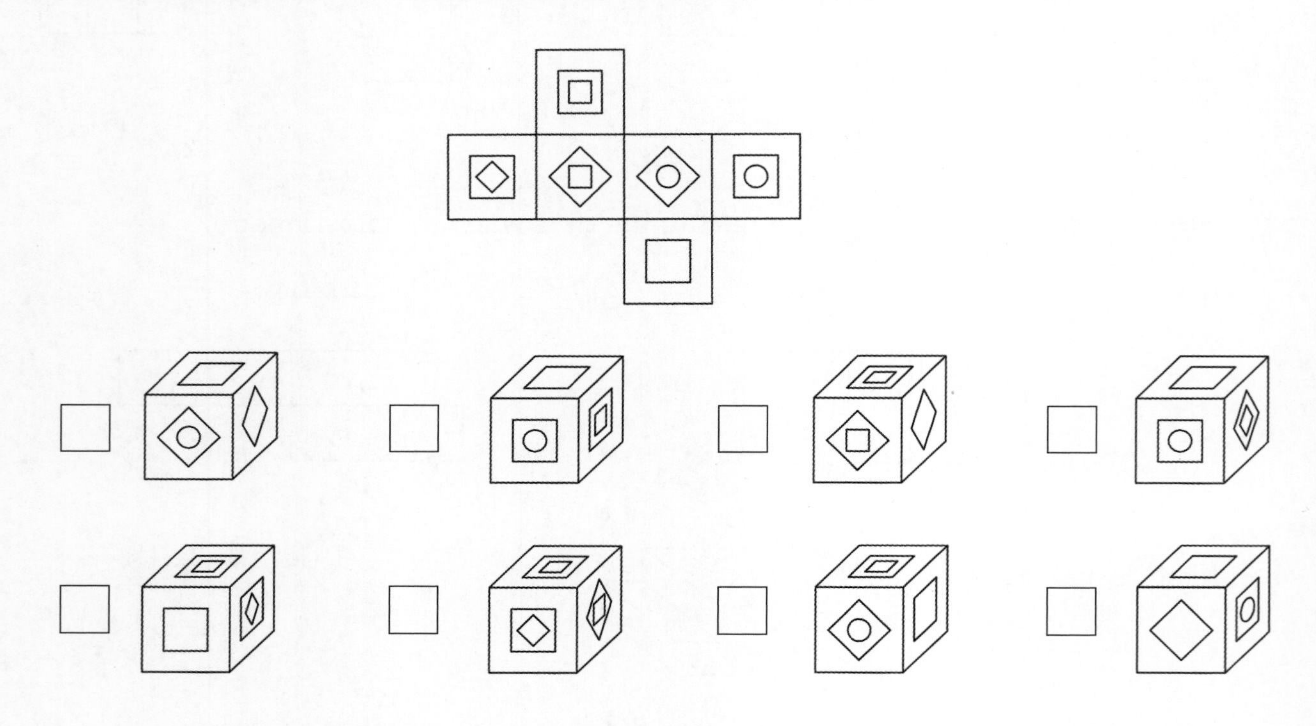

7 These nets are for an open cylinder. Which one will form a cylinder in which the symbols match?

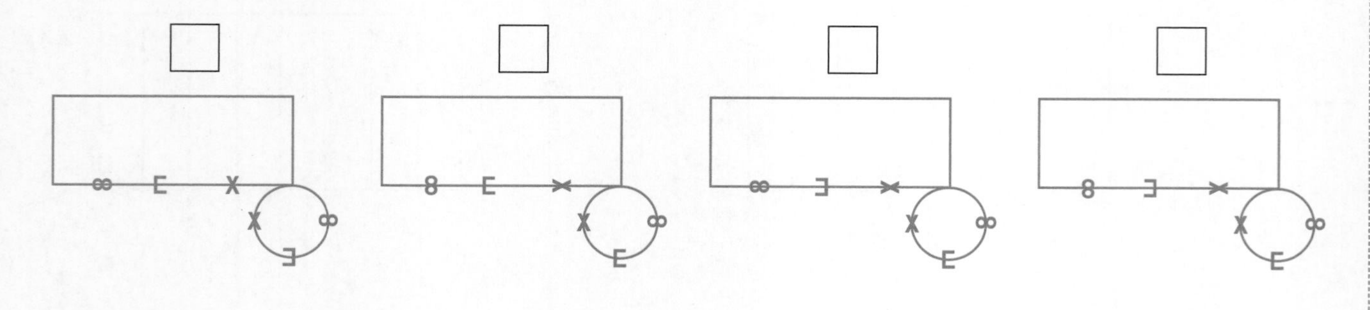

 ISBN: 9780170477710

N Location and navigation

Directions

- Directions can be given in terms of north, south, east and west, or combinations of these.

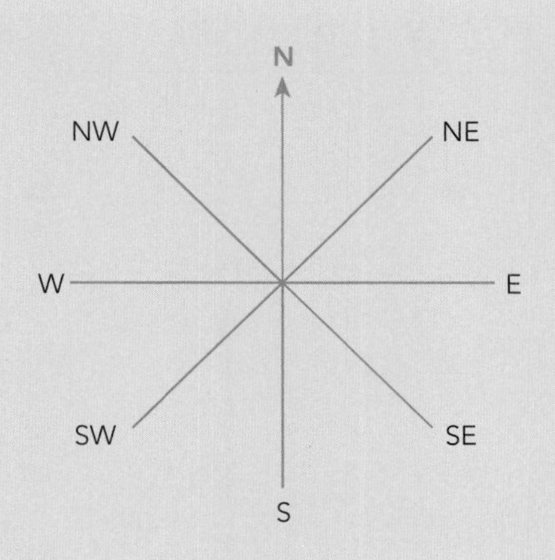

Examples:

1 Mia is looking at some house plans.

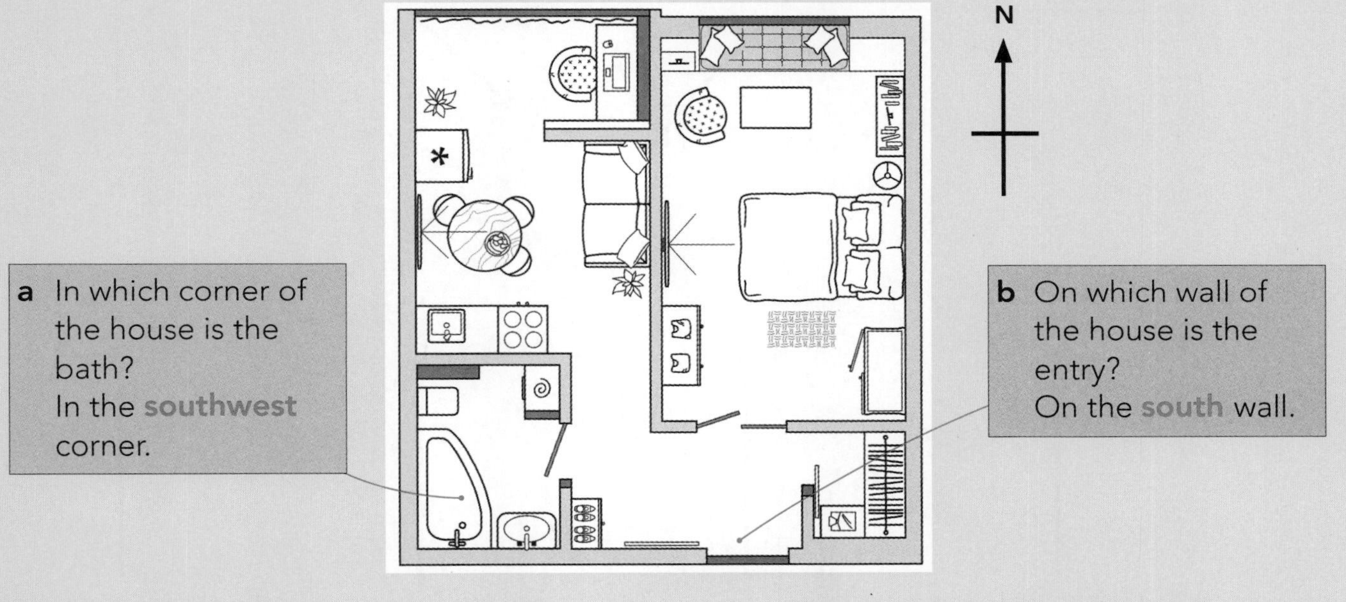

a In which corner of the house is the bath?
In the **southwest** corner.

b On which wall of the house is the entry?
On the **south** wall.

2 Adam is driving from Red Post Corner to Rotherham. In which direction is he driving?

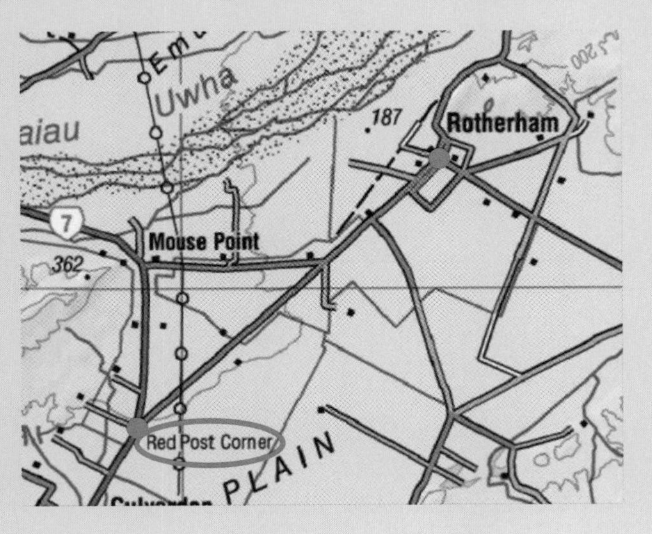

He is driving **NE (northeast)**.

ISBN: 9780170477710

N

Answer the following questions.

1 Match the compass directions of the green lines to those in the list. They are either in the direction of the arrow, or from A to B. Use each direction once only.

W	NE	S	E	NW	SE	SW	N

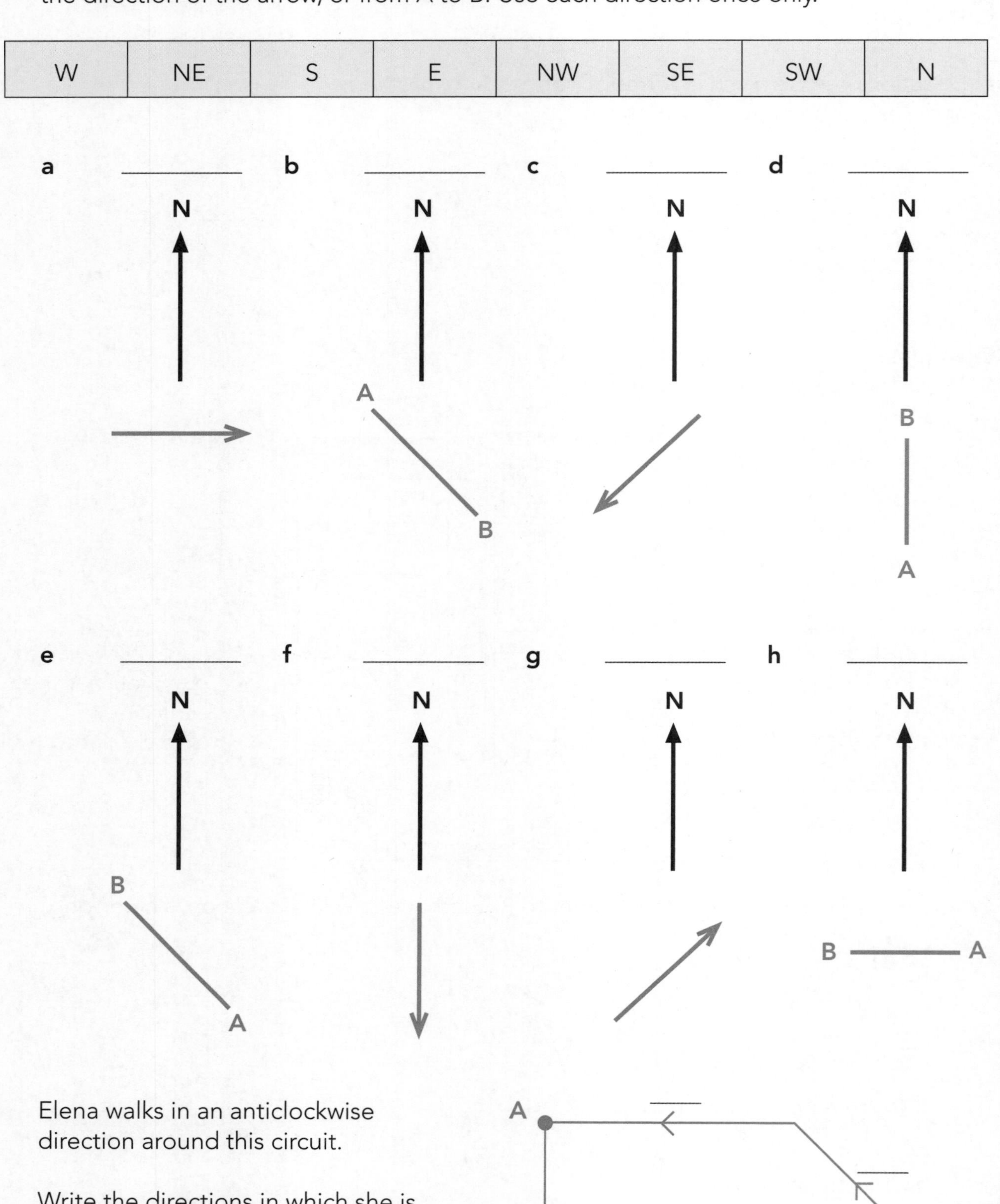

2 Elena walks in an anticlockwise direction around this circuit.

Write the directions in which she is walking on the lines provided.

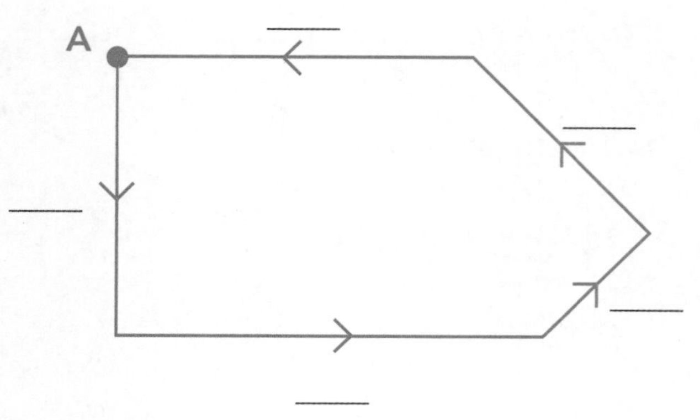

ISBN: 9780170477710

3 **a** Nikora lives in Gladstone. On Saturday he drove to Masterton to do his shopping. In which direction is Masterton from Gladstone?

______

b After shopping in Masterton, he drove to see his koro in Carterton. In which direction did he drive?

______

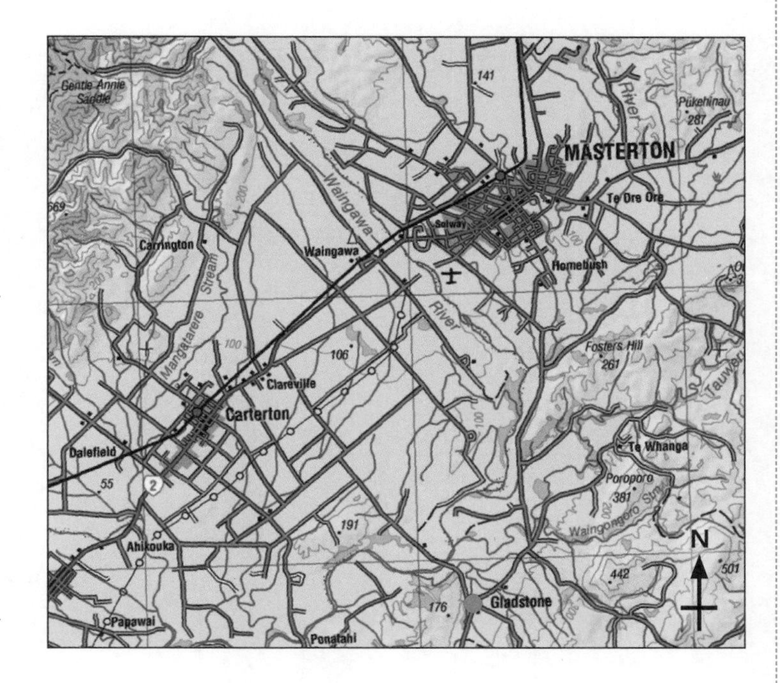

4 The map shows the township of Palmerston in Otago.

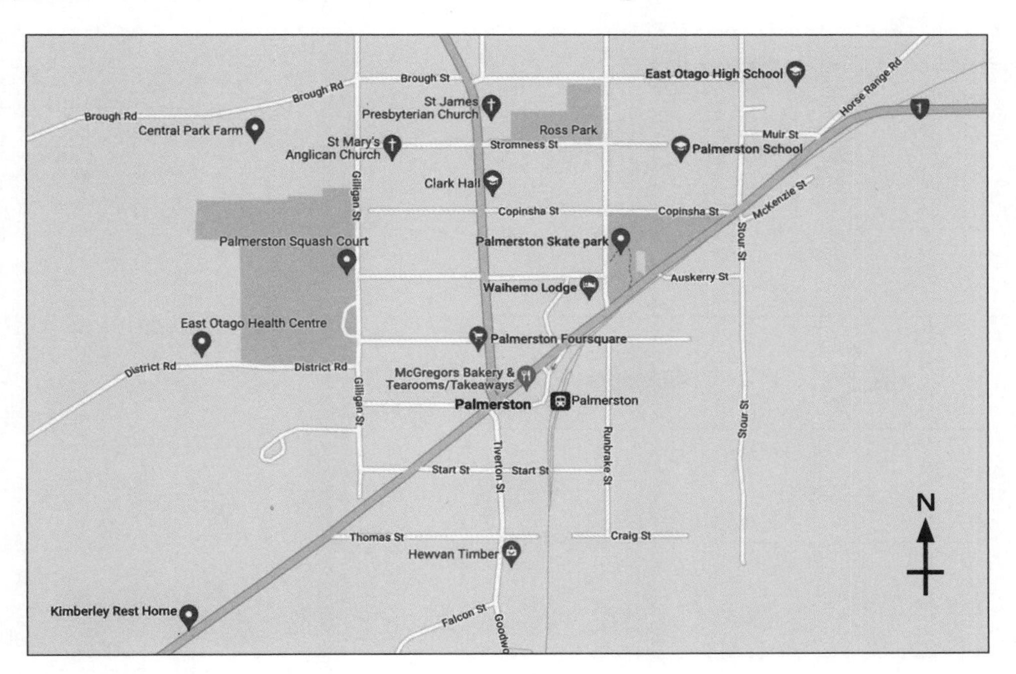

a Maia has two part-time jobs. In the mornings she works at the Kimberley Rest Home, and in the afternoons she does cleaning at Waihemo Lodge. In which directions does she need to drive at lunchtime?

______

b A visitor to the Kimberley Rest Home asks her for directions to St James Presbytarian Chruch. Fill in the gaps.

Turn left out of Kimberley Rest Home and drive in a ______ direction to the town.

Then turn ______ and drive in a ______ direction for four and half blocks.

ISBN: 9780170477710

Location: grid references

- Grid references are used to identify **locations** on a map.
- Locations are specified by **two** coordinates.
- As with coordinates on graphs, **horizontal** coordinates are always given **before vertical** coordinates.

Locating squares

The kangaroo:

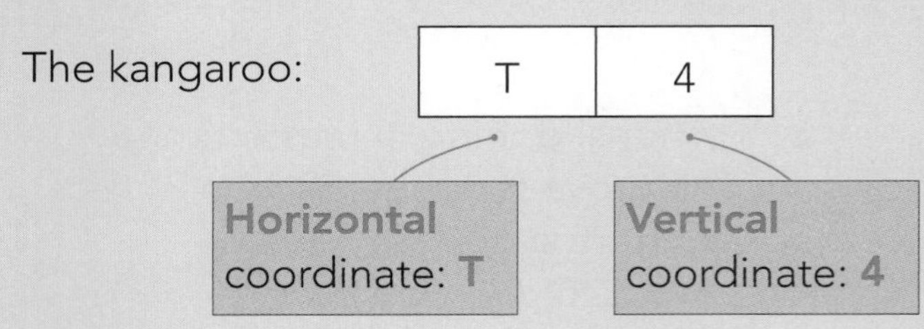

Check for yourself:

The pig:	U	2
The penguin:	S	1
The owl:	R	3

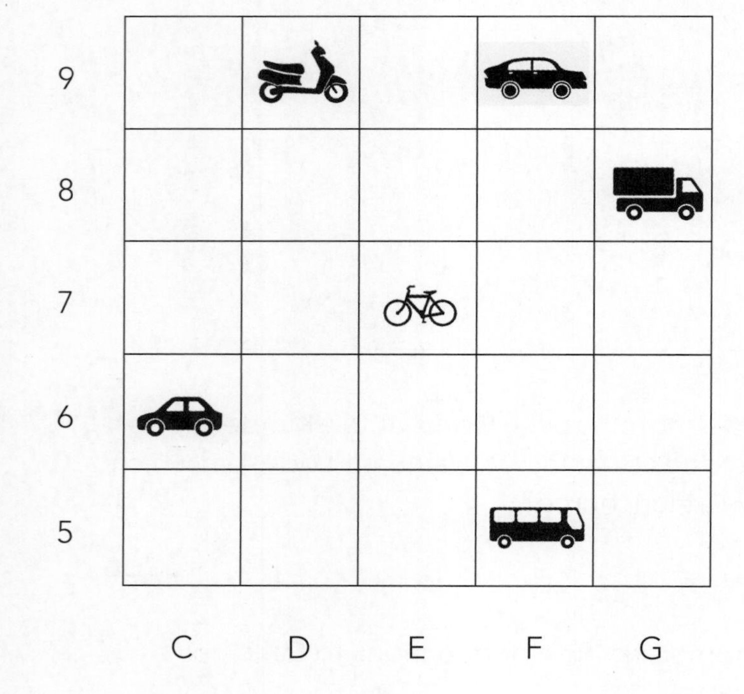

What would you find in the following squares?

1 F5 ____________________

2 E7 ____________________

Write coordinates for the squares where you would find the following.

3 Motor scooter __________

4 Truck __________

 ISBN: 9780170477710

This map shows a section of Orana Wildlife Park.

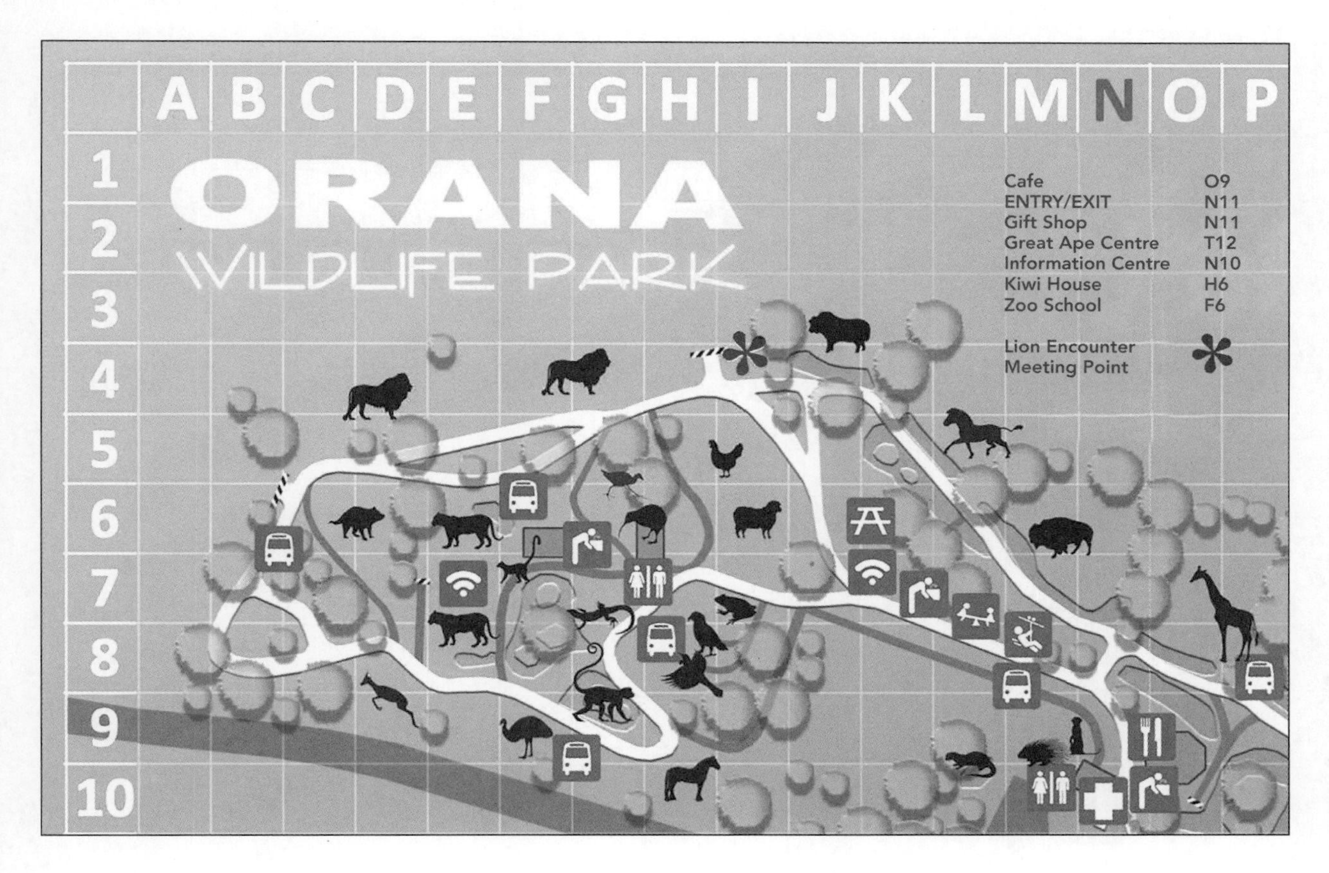

What would you find in the following squares?

5 E6 ____________________

6 I4 ____________________

7 K6 ____________________

8 H10 ____________________

Write grid references for the squares containing the following locations.

9 Frogs __________

10 Emu __________

11 Sheep __________

12 First Aid __________

ISBN: 9780170477710

N

Locating grid points

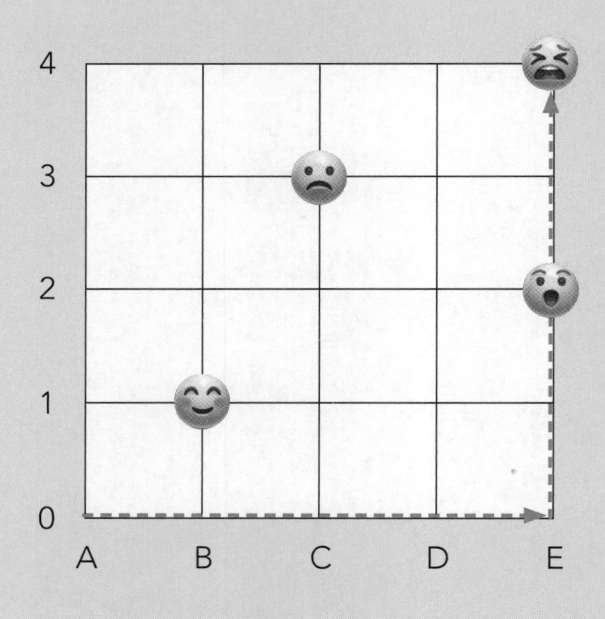

The crying face:

E	4

Horizontal coordinate: E

Vertical coordinate: 4

Check for yourself:

The smiley face:

B	1

The surprised face:

E	2

The sad face:

C	3

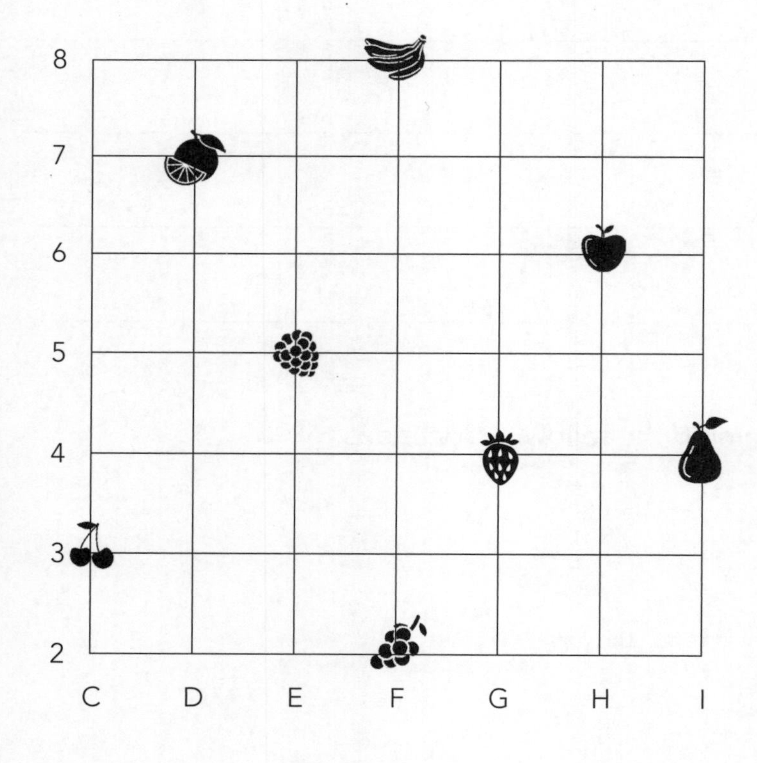

What would you find at the following grid points?

1 F8 ____________________

2 C3 ____________________

3 H6 ____________________

4 G4 ____________________

Write coordinates for the grid points where you would find the following.

5 Orange __________

6 Pear __________

7 Grapes __________

8 Raspberry __________

 ISBN: 9780170477710

Scales

- A scale gives a measure of **distance** on a map or diagram.
- You will not be able to determine exact distances, but you need to be able to **estimate** distances.
- Estimations should be made to no more than two significant figures.

Examples:

1 The distance between the gridlines is 10 km.
Estimate the distance between Culverden and Rotherham.

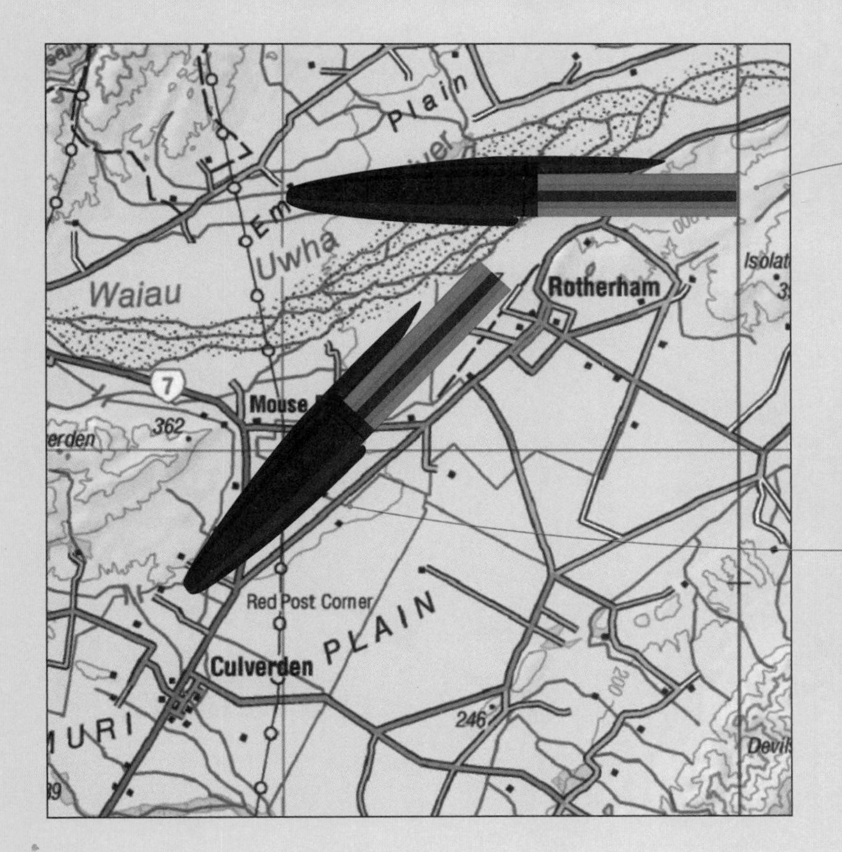

Use part of a pen or your finger to measure the distance between the gridlines.

Then use the pen to estimate the distance needed.
In this case the distance between Culverden and Rotherham is a little more than one part-pen length, so a good estimate would be 12 km.

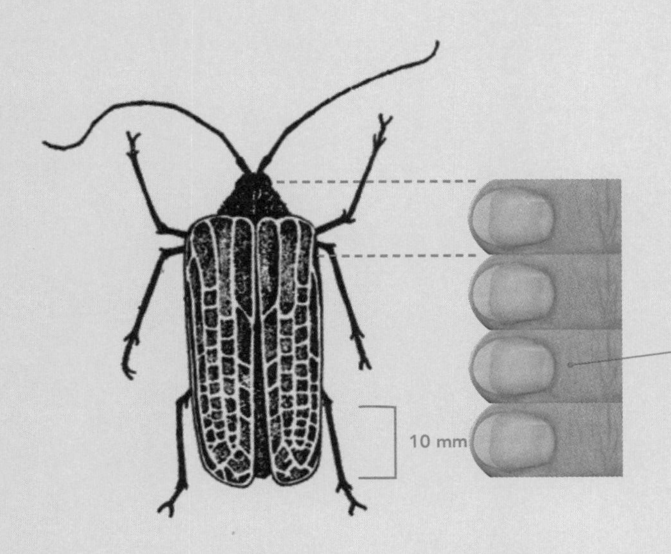

You could use the width of a fingernail to estimate 10 mm, and then count how many make up the length of a wing. The wings are between three and four fingernails long (30–40 mm), so a good estimate would be 35 mm.

ISBN: 9780170477710

N

Answer the following questions.

1 This map shows an area in South Canterbury. The distance between the gridlines is 10 km. Estimate the following direct distances.

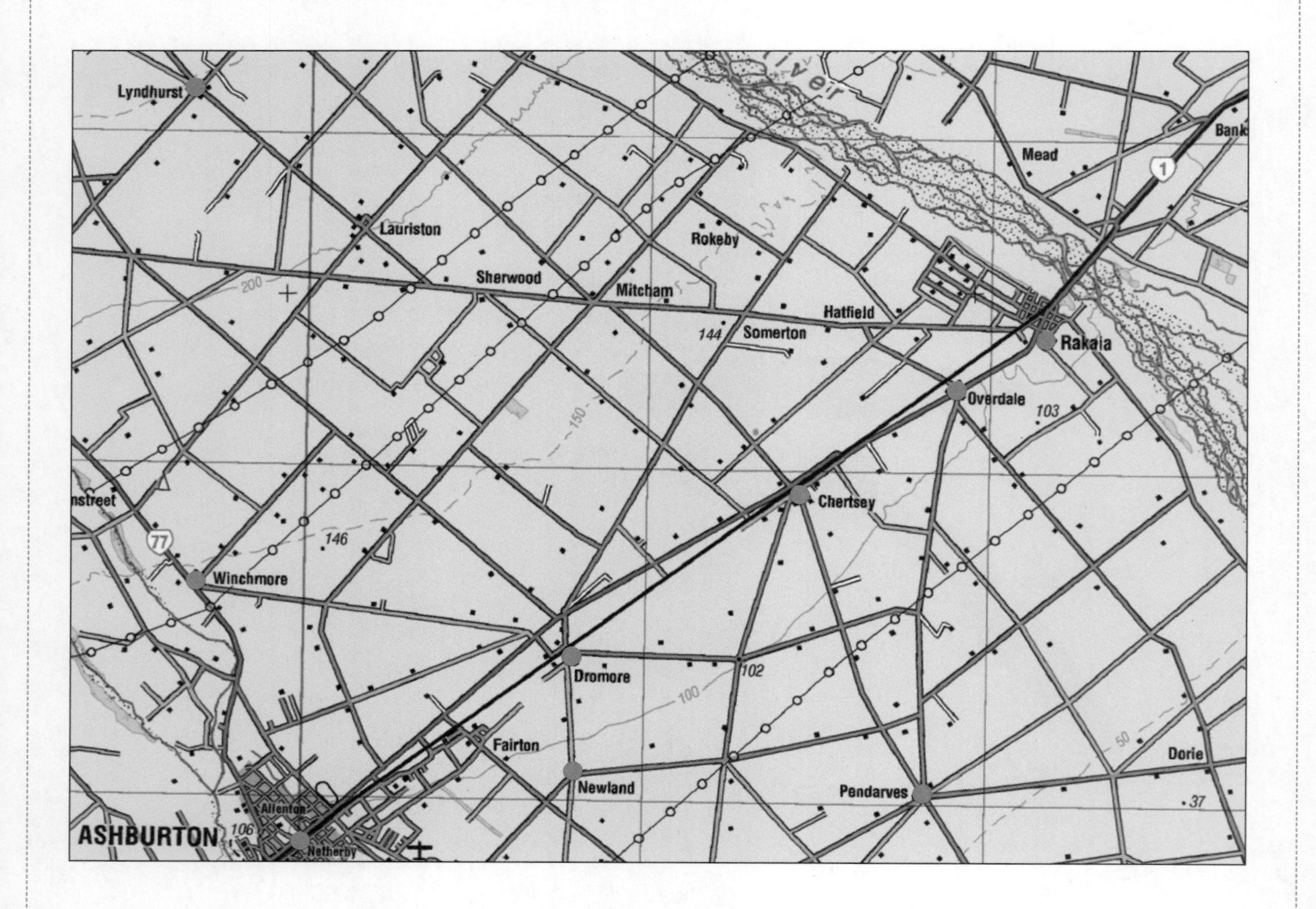

a Pendarves to Chertsey. _______________ km

b Pendarves to Overdale. _______________ km

c Newland to Dromore. _______________ km

d Winchmore to Dromore. _______________ km

e Ashburton to Rakaia. _______________ km

f Winchmore to Chertsey. _______________ km

g Lyndhurst to Rakaia. _______________ km

ISBN: 9780170477710

2 This is the plan for a house. The lines on the grid are 1 metre apart. Use a ruler to help you to find the following.

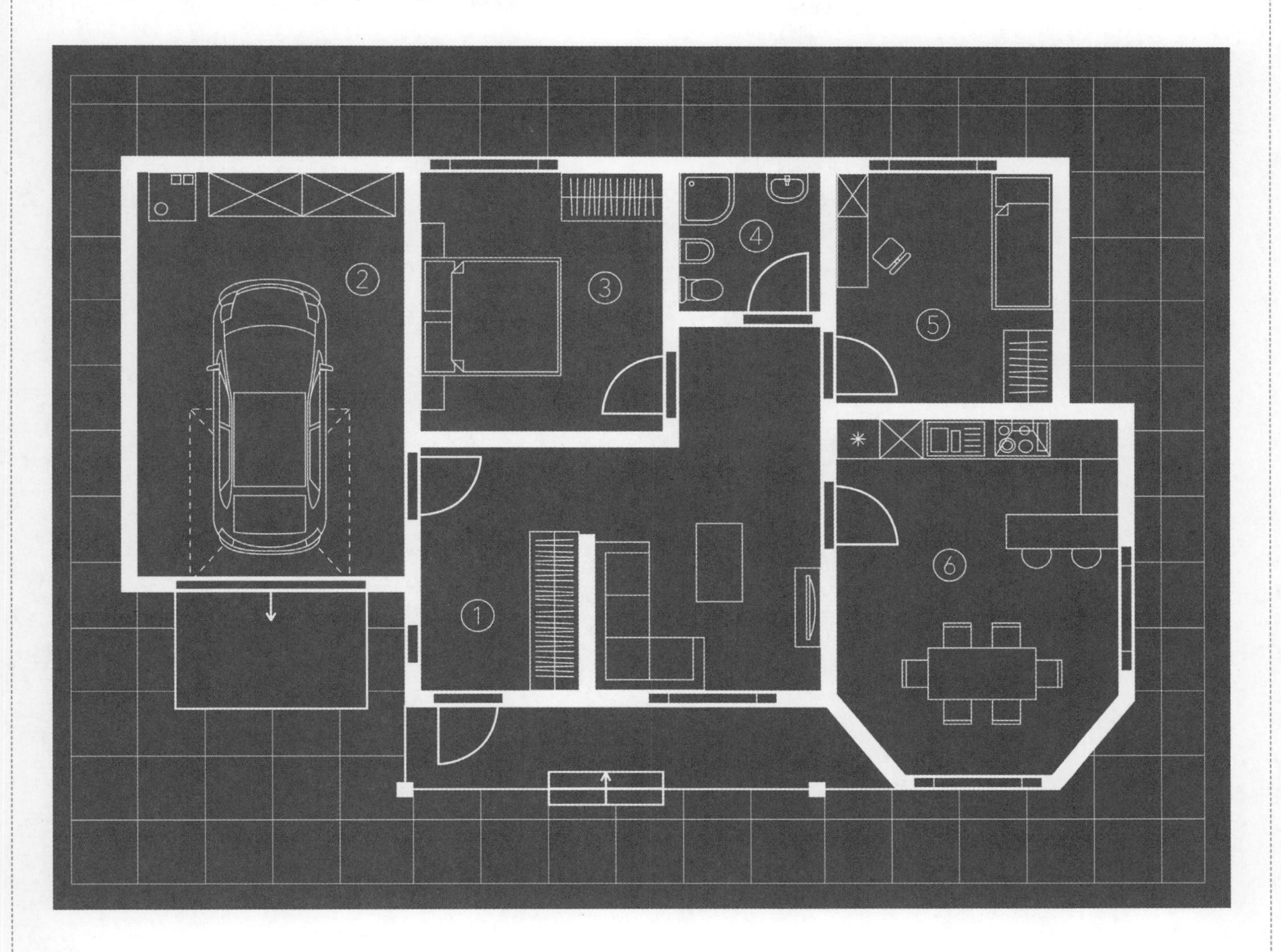

a The length of the car. ______________ m

b The internal length of the garage. ______________ m

c The internal width of the garage. ______________ m

d The width of the double bed. ______________ m

e The length of the dining table. ______________ m

f The internal length of one wall of the square bathroom. ______________ m

g The total width of the house ______________ m

ISBN: 9780170477710

N

3 This photograph shows a koura, a freshwater crayfish.

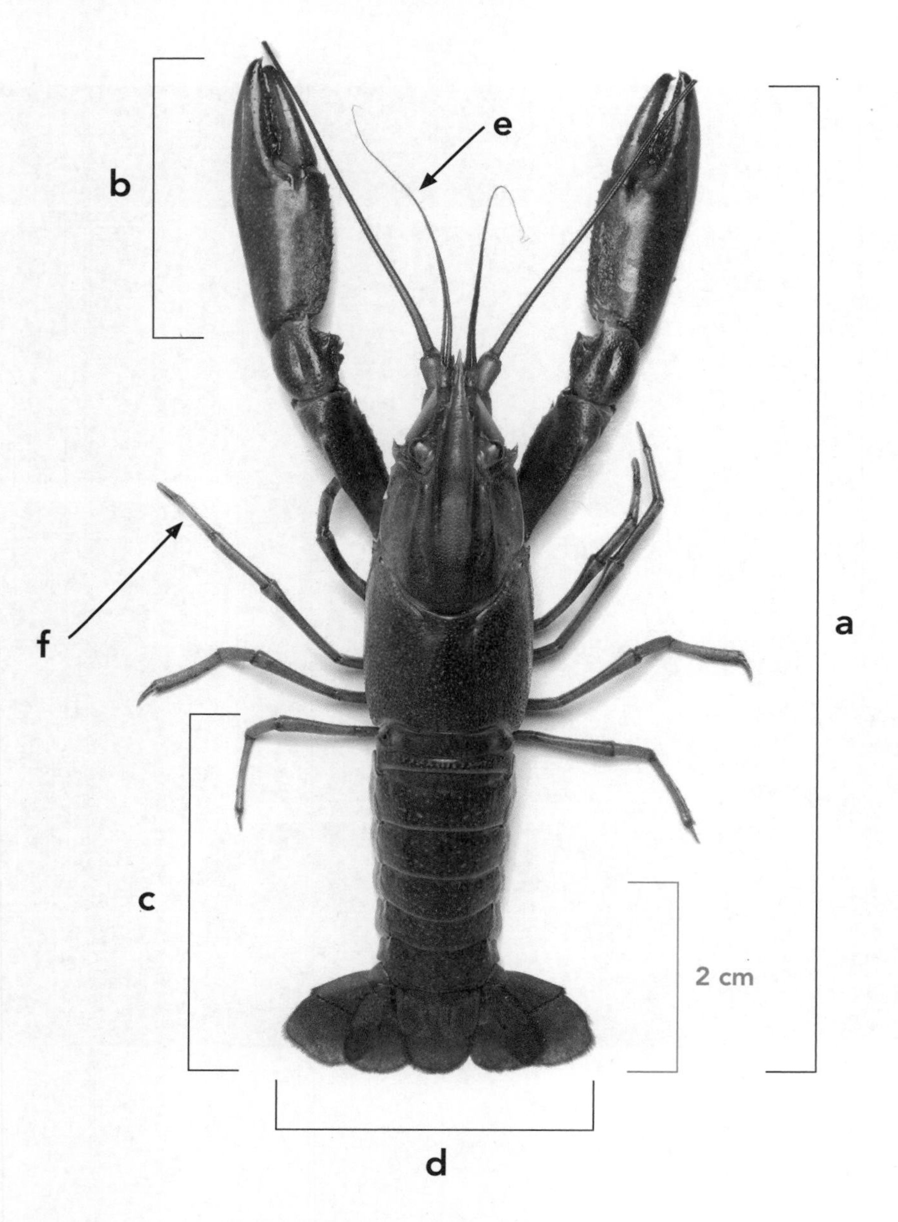

a Estimate the total length of this koura to the nearest centimetre. ______ cm

Estimate the following lengths to the nearest half centimetre.

b The length of the claw. ______ cm

c The total length of the tail. ______ cm

d The width of the tail. ______ cm

e The length of this feeler. ______ cm

f The length of this leg. ______ cm

ISBN: 9780170477710

Mixing it up

The map shows some of the major centres in Asia and the pacific. The distance between the gridlines is 2000 km.

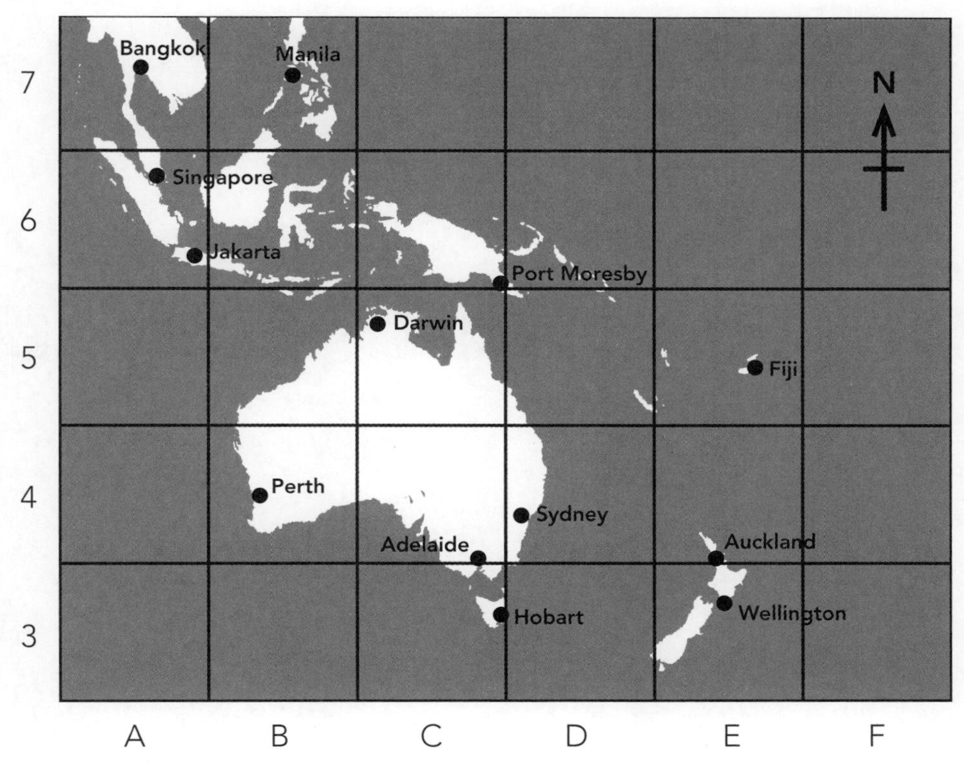

1 Which centres lie in the following squares?

a E5 ______________ **b** B7 ______________

2 Which squares do the following centres lie in?

a Bangkok ______________ **b** Perth ______________

3 In which direction would you be flying the following routes?

a Auckland to Adelaide: ________ **b** Hobart to Port Moresby: ________

c Jakarta to Adelaide: ________ **d** Sydney to Singapore: ________

e Fiji to Sydney: ________ **f** Singapore to Manila: ________

4 Estimate the distance between the following centres to the nearest 100 km.

a Darwin and Port Moresby: ______ km **b** Auckland and Wellington: ______ km

c Darwin and Perth: ______ km **d** Singapore and Manila: ______ km

e Fiji and Port Moresby: ______ km **f** Auckland and Bangkok: ______ km

ISBN: 9780170477710

N Measurement

Time

Converting units

- Time can be measured in seconds, minutes, hours, days, etc.

Use the following chart to help you convert times.

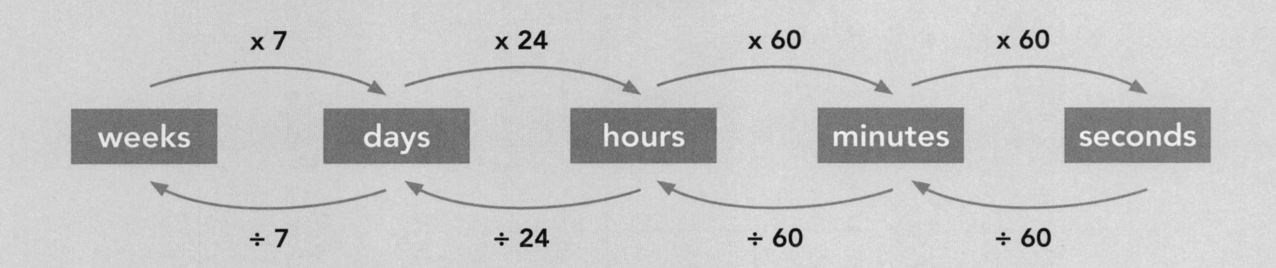

You also need to know: 1 that there are about 4 weeks in a month.
2 the names and order of the 12 months in a year.

Example:

How many seconds are in 9 minutes?

9 min = 9 x 60 seconds
= 540 seconds

There are 60 seconds in a minute.

1 Tick the correct calculation for each of these conversions.

a 480 hours into days

☐ 480 ÷ 24 ☐ 480 x 60 ☐ 480 x 24 ☐ 480 ÷ 60

b 5 days into minutes

☐ 5 x 12 x 60 ☐ 5 x 24 x 60 ☐ 5 x 24 x 60 x 60 ☐ 5 x 12 x 60 x 60

c 504 hours into weeks

☐ 504 ÷ 12 ÷ 7 ☐ 504 x 24 ÷ 7 ☐ 504 ÷ 12 x 7 ☐ 504 ÷ 24 ÷ 7

2 Highlight the correct conversion for each of the following.

a	120 s	1 min	2 min
		2 h	1.5 min
b	30 h	360 min	108 000 min
		180 min	1800 min
c	5 weeks	2100 min	300 min
		126 000 min	50 400 min
d	2 days	24 h	2880 min
		36 h	2800 min

ISBN: 9780170477710

3 Convert the following.

a 240 s = _______________ min

b 2 days = _______________ h

c 360 min = _______________ h

d 2.5 h = _______________ min

e 120 min = _______________ s

f 180 h = _______________ days

g 3 weeks = _______________ h

h 3 days = _______________ min

4 Highlight the longer period of time.

a 1.5 h   100 min

b 1.5 days   35 h

c 190 s   3 min

d 170 min   3 h

e 4 h   245 min

f 7.5 min   460 s

g 3700 min   2.5 days

h 2 days   172 801 s

5 Convert these times to minutes and then put them in order from shortest to longest.

a

95 min	0.1 day = _______	6000 s = _______	1.6 h = _______

Shortest   Longest

b

1.4 day = _______	150 000 s = _______	30 h = _______	2000 min

Shortest   Longest

N

12- and 24-hour time

There are two ways of writing times:

12-hour time can also be written with a full stop, e.g. 1.00 am.

24-hour time can also be written without the colon, e.g. 0000.

Midnight or 12:00 am.

12-hour time	24-hour time
Midnight	00:00
1:00 am	01:00
2:00 am	02:00
3:00 am	03:00
4:00 am	04:00
5:00 am	05:00
6:00 am	06:00
7:00 am	07:00
8:00 am	08:00
9:00 am	09:00
10:00 am	10:00
11:00 am	11:00
Midday or noon	12:00
1:00 pm	13:00
2:00 pm	14:00
3:00 pm	15:00
4:00 pm	16:00
5:00 pm	17:00
6:00 pm	18:00
7:00 pm	19:00
8:00 pm	20:00
9:00 pm	21:00
10:00 pm	22:00
11:00 pm	23:00

am tells you that a time is in the **morning**.

24-hour time has **four digits**, so a **0** is added to the start of times earlier than 10 am.

Midday or noon or 12:00 pm.

pm tells you that a time is in the **afternoon or evening**.

Between 1 pm. and 11:59 pm: To convert 12-hour times into 24-hour time, you need to add **12 hours**.

 ISBN: 9780170477710

6 Highlight the correct conversion for each of the following.

a	3.29 pm	03:29	15:29	13:29
b	7:45 am	31:45	19:45	07:45
c	11:49 pm	23:49	11:49	01:49
d	12:01 am	24:01	00:01	12:01
e	22:30	10:30 am	10:30 pm	22:30 pm
f	13:55	1:55 am	13:55 pm	1:55 pm
g	04:09	4:09 am	04:09 am	04:09 pm
h	12:38	0:38 am	0:38 pm	12:38 pm

7 Convert the following to 24-hour time.

a 9:08 pm = ______________

b 3:04 am = ______________

c 11:59 pm = ______________

d 10:45 am = ______________

e Midday = ______________

f Midnight = ______________

8 Convert the following to 12-hour time.

a 15:11 = ______________

b 18:50 = ______________

c 07:42 = ______________

d 22:27 = ______________

e 12:48 = ______________

f 00:23 = ______________

9 Highlight the time that is later in the day.

a 3:45 am 04:35

b 3:52 pm 13:53

c 6:21 pm 18:12

d 13:01 2.02 pm

e 11:21 am 22:20

f 10:00 pm 00:00

N

Time intervals

Hours and minutes

Examples:

1 A 1½-hour exam started at 2:45 pm. When should it finish?

```
  2 hours + 45 min
+ 1 hour  + 30 min
  3 hours + 75 min = 4.15 pm
```

2 It will take Bert 2 hours and 40 minutes to drive to an away match. He needs to start his warm-up at 1:30 pm. What is the latest time that he can leave?

1:30 pm = 13:30.

It's often easiest to work in 24-hour time.

```
  13:30
– 02:40
```

Not easy to subtract.

But 13:30 = 12:30 + 60 min = 12:90.

```
  12:90
– 02:40
  10:50
```

Easier to subtract.

So the latest Bert can leave is 10:50 am.

10 Find the following times. Space has been left for your working on the right. Use the same 'style' for writing the answers as in the question.

a 5½ hours after 01:30 = __________

b 25 minutes after 12:45 = __________

c 50 minutes before 00:30 = __________

d 25 minutes before 13:10 = __________

e 13 hours 40 minutes after 05:30 = __________

f 6½ hours before 23:15 = __________

g 2½ hours after 11:30 am = __________

h 7 hours and 10 minutes before 5:45 pm = __________

i 2 hours and 50 minutes before 1:25 am = __________

ISBN: 9780170477710

11 Find the differences between these times.

a 01:17 and 02:42 = ______ h _______ min

b 13:01 and 16:42 = ______ h _______ min

c 03:41 and 05:36 = ______ h _______ min

d 6:21 am and 10:52 pm = ______ h _______ min

e 09:37 and 23:09 = ______ h _______ min

f 7:34 am and 11:13 am = ______ h _______ min

Weeks, months and years

Remember: 1 There are about 30 days or about 4 weeks in each month.

2 The 12 months in a year in order are:

January	February	March	April	May	June
July	August	September	October	November	December

12 a 8 weeks after 18 June will be early/mid/late ______________________

b 10 weeks after 1 March will be early/mid/late ______________________

c 6 weeks before 30 September will be early/mid/late ______________________

d 12 weeks before 14 December will be early/mid/late ______________________

e 5 months after May 2022 will be ______________________ 20 ________

f 18 months after July 2022 will be ______________________ 20 ________

g 7 months before February 2024 will be ______________________ 20 ________

h 15 months before June 2024 will be ______________________ 20 ________

N

Reading timetables

- To use timetables, you need to be able to:
  - understand 12- and 24-hour time
  - calculate differences between times.
- Timetables can be vertical or horizontal.

Examples:

1

This bus will be at the Kelly Place stop at 09:13. Then will arrive at The Terrace at 09:41.

09:36 – 09:13 = 23 minutes. So the next bus to leave Kelly Place will be 23 minutes later.

Kelly Place	09:13	09:36
Manuka Road	09:27	09:50
The Terrace	09:41	10:04

2

Totara Lane	Tui Street	Valley Road
10:58 am	11:06 am	11:15 am
11:21 am	11:29 am	11:38 am
11:44 am	11:52 am	12:01 pm
12:07 pm	12:15 pm	12:24 pm

This bus will be at Totara Lane at 10:58 am. It will reach Valley Road at 11:15 am.

12:01 pm – 11:52 am = 9 minutes. The trip from Tui Street to Valley Road takes 9 minutes.

Answer the following questions.

13 Bus timetable

a How often do buses leave from Gibbon Drive?

Every ______________ min

b How long is the bus ride from Essex Avenue to King Street?

______________ min

c Ronan needs to be at Bulwer Place before 8 am. What time at the latest should he get the bus from Ajax Road?

____

Ajax Road	07:03	07:18	07:33	07:48
Gibbon Drive	07:08	07:23	07:38	07:53
Essex Avenue	07:17	07:32	07:47	08:02
Bulwer Place	07:24	07:39	07:54	08:09
Sandy Drive	07:35	07:50	08:05	08:20
King Street	07:43	07:58	08:13	08:28

ISBN: 9780170477710

14 Train timetable

Greenville	Avalon	Warrington	Birkdale	Long Beach	Cavern Bay
10:55 am	11:03 am	11:09 am	11:20 am	11:32 am	11:46 am
11:15 am	11:23 am	11:29 am	11:40 am	11:52 am	12:06 pm
11:35 am	11:43 am	11:49 am	12:00 pm	12:12 pm	12:26 pm
11:55 am	12:03 pm	12:09 pm	12:20 pm	12:32 pm	12:46 pm
12:15 pm	12:23 pm	12:29 pm	12:40 pm	12:52 pm	1:06 pm
12:35 pm	12:43 pm	12:49 pm	1:00 pm	1:12 pm	1:26 pm

a How often does a train leave from Greenville? Every ______ min

b How long does the trip from Birkdale to Cavern Bay take? ______ min

c Greta needs to be at Cavern Bay by midday. Which train should she take from Avalon? ______

d How long does the trip from Greenville to Long Beach take? ______ min

e Wiremu is taking a 13:12 train. Which station is it leaving from? ______

15 Ferry timetable

Waima	Titoki	South Bay	Tawa	Poto Beach
13:08	13:29	13:57	14:21	14:40
13:37	13:58	14:26	14:50	15:09
14:03	14:24	14:52	15:36	15:55
14:48	15:09	15:37	16:01	16:20
15:14	15:35	16:03	16:27	16:46
15:51	16:32	17:00	17:24	17:43
16:25	16:46	17:34	17:58	18:37
16:48	17:09	17:37	18:01	18:20

a How long does it take the ferry to get from Waima to South Bay?

______ min

b Tia takes the 2:24 pm from Titoki. At what time (in 12-hour time) will she arrive at Poto Beach?

______

c Jessie arrives at the Titoki ferry terminal at 4:30 pm. What time will the next ferry leave?

______

d Yvonne wants to be at Tawa by 6 pm. Which ferry should she take from Waima?

______

N

Scales

Reading scales

- Find **zero** on the scale.
- Make sure that you read in the **correct direction**.
- If zero is not on the scale, make sure you read from **smaller values to bigger values**.
- Include **units** in you answer.
- Think about your answer. **Does it seem reasonable?**

Write down the measurements shown on these scales. Include units in your answers.

1

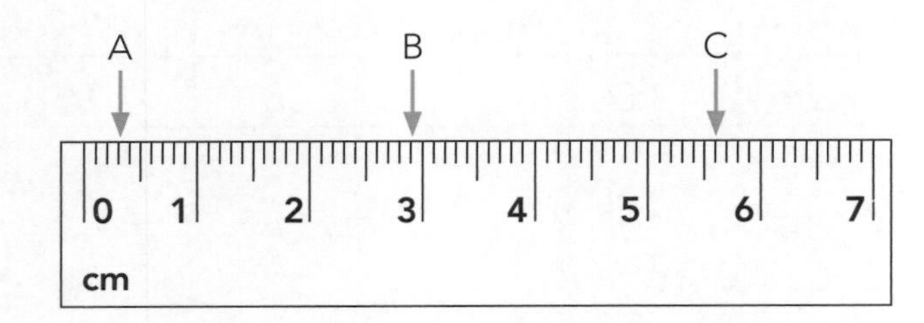

A = ______________ cm

B = ______________ cm

C = ______________ cm

2

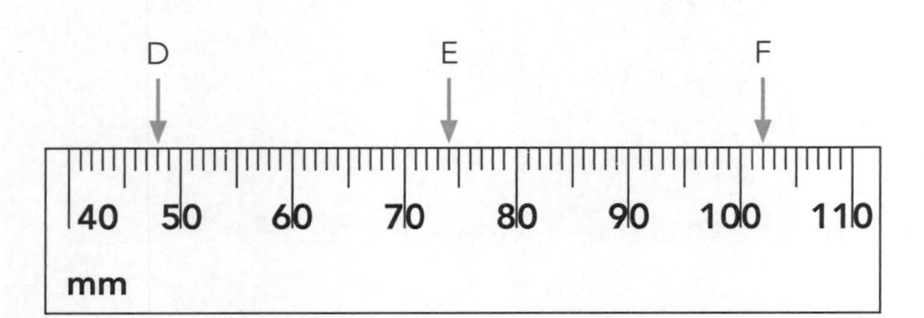

D = ______________ mm

E = ______________ mm

F = ______________ mm

3

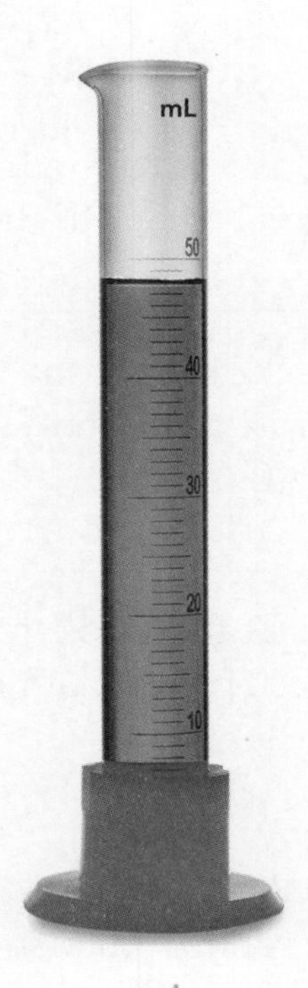

Volume ________ mL

4

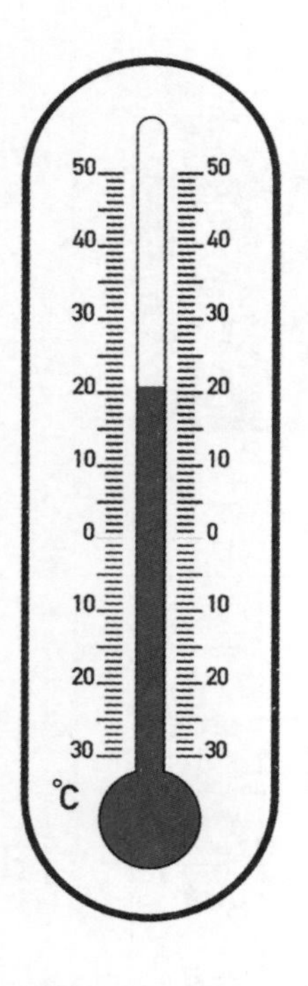

Temperature ________ °C

5

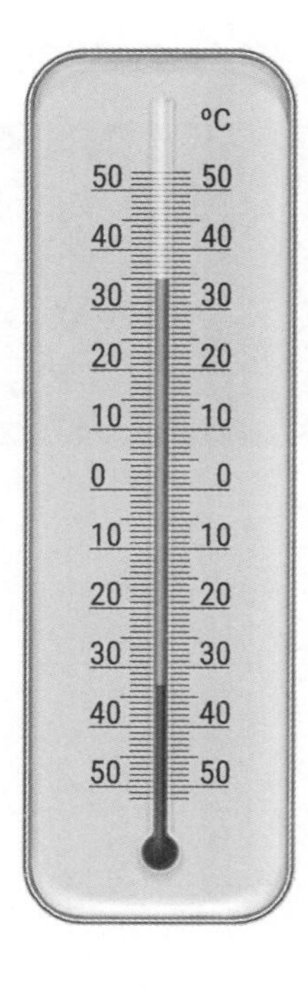

Temperature ________ °C
(Be careful)

 ISBN: 9780170477710

6

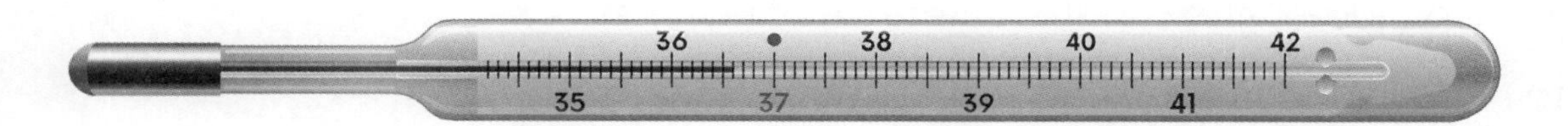

Temperature _______ °C

7

Speed _______ km/h

8

Mass _______ kg

9

Current _______ amps

10

Voltage _______ volts

11 This is part of a 5 metre measuring tape. Write the measure in metres. _______ m

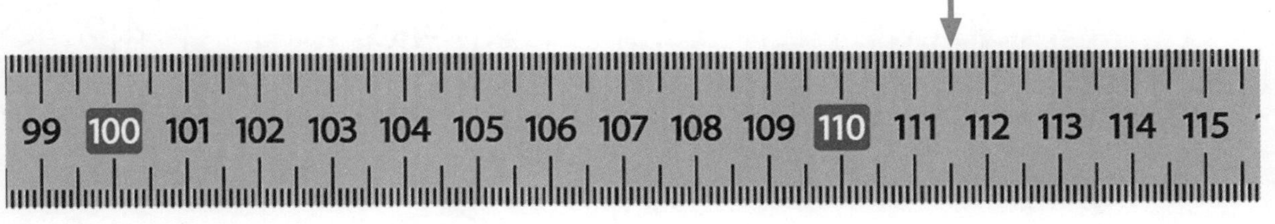

ISBN: 9780170477710

N

Showing values on scales

Colour or add an arrow to show these measurements.

12 180 mL

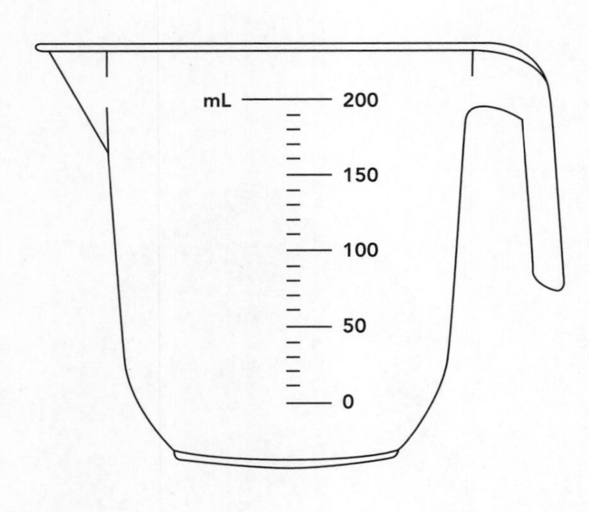

13 95 km/h

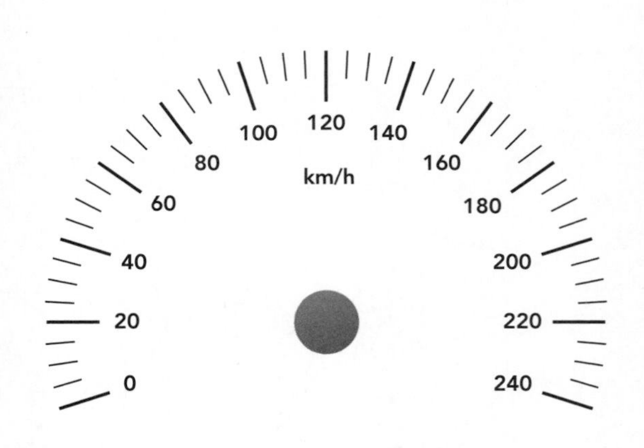

14

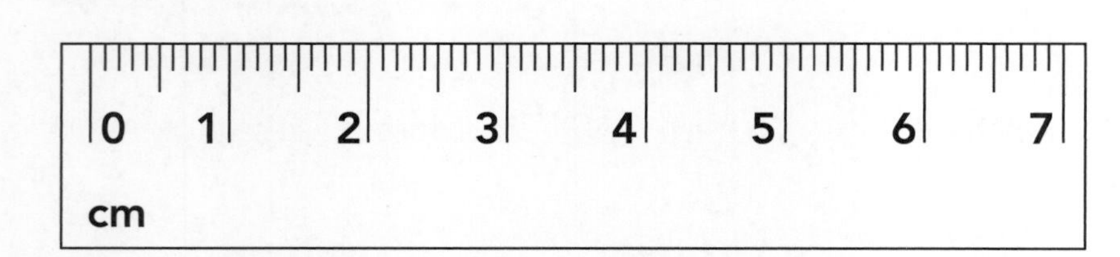

A = 4.6 cm

B = 6.8 cm

C = 0.3 cm

15 88 kg

16 ¾ of a tank

17 40.3°C

ISBN: 9780170477710

18 49 V

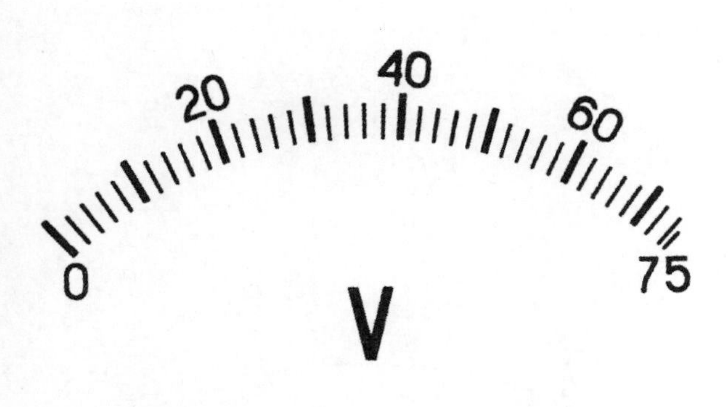

19 51.5 Hz

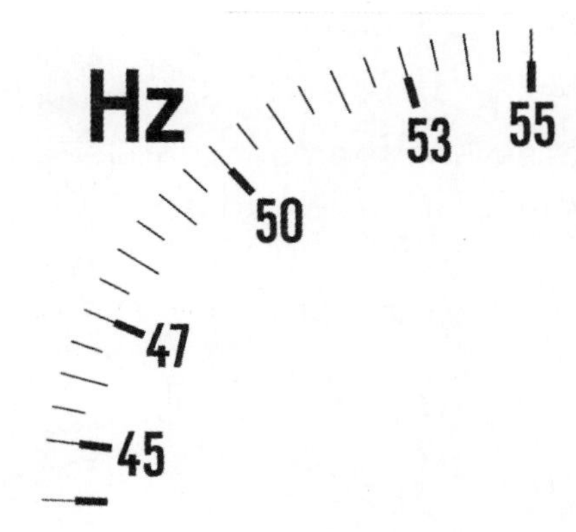

20 6 kg

21 Write decimals values for each point along the ruler. Choose the most appropriate values from the options below. You will not need all of them.

3.0	6.2	2.1	7.8	0.2	4.9
1.4	4.3	2.2	5.6	7.3	7.5
6.4	0.7	7.1	3.8	4.5	2.5

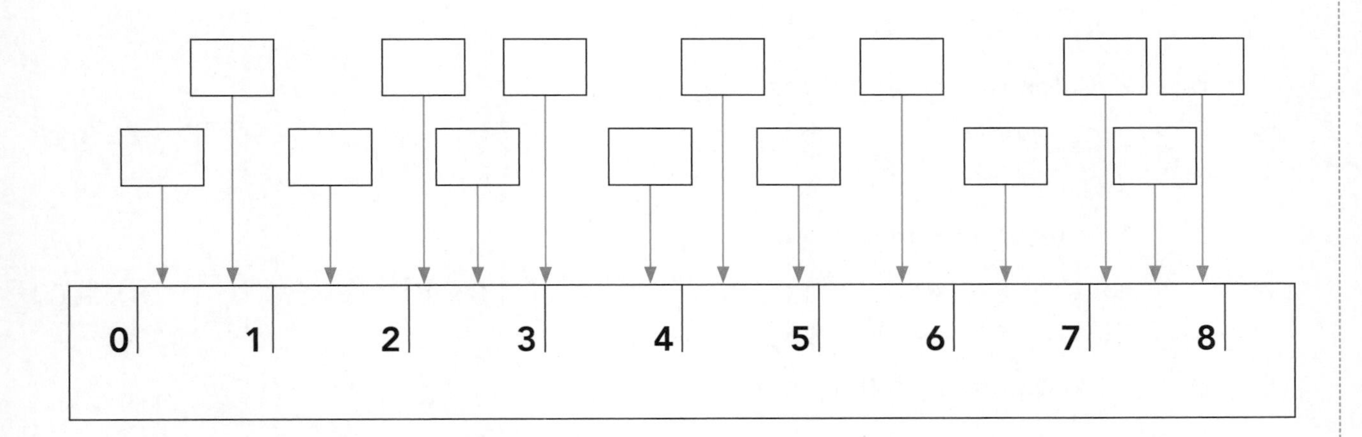

ISBN: 9780170477710

N

Converting units

Length

- The basic unit for measuring length is the **metre**.
- All other units of length in the metric system are based on the metre.

Use the following chart to help you convert lengths.

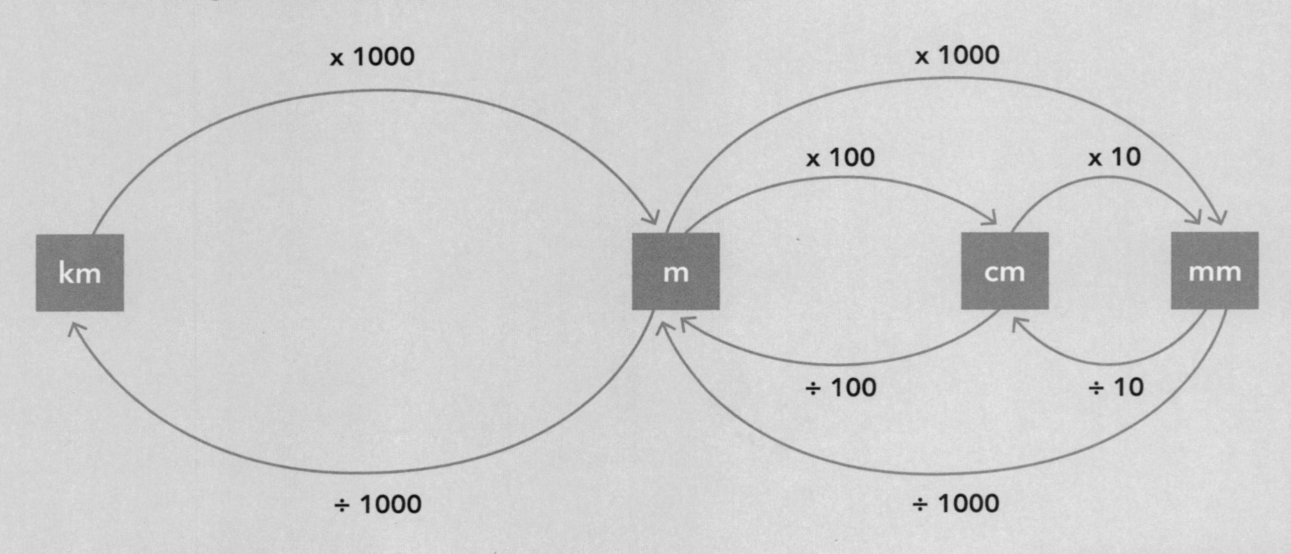

Example:

1 How many metres are in 1.6 kilometres?

1.6 km = 1.6 x 1000 m
= 1600 m

There are 1000 m in 1 km.

2 How many metres are in 930 centimetres?

930 cm = (930 ÷ 1000) cm
= 9.3 m

There are 100 cm in 1 m.

1 Tick the correct calculation for these conversions.

a 500 centimetres into metres

☐ 500 x 100 ☐ 500 ÷ 100 ☐ 500 ÷ 10 ☐ 500 x 10

b 2.7 kilometres into metres

☐ 2.7 ÷ 1000 ☐ 2.7 x 1000 ☐ 2.7 x 100 ☐ 2.7 ÷ 100

c 1.8 metres into millimetres

☐ 1.8 ÷ 10 x 1000 ☐ 1.8 x 100 ☐ 1.8 x 10 x 10 ☐ 1.8 x 1000

d 0.6 kilometres into millimetres

☐ 0.6 x 100 x 100 ☐ 0.6 x 1000 x 10 ☐ 0.6 x 1000 x 1000 ☐ 0.6 x 1000

ISBN: 9780170477710

Highlight the correct conversion for each of the following.

2	6000 mm	0.6 m	600 cm
		0.06 km	6 cm
4	2.1 km	21 000 cm	210 m
		21 m	2100 m
6	5.3 m	5300 cm	0.53 km
		0.053 km	5300 mm

3	0.7 m	7000 mm	70 cm
		7 km	700 cm
5	850 cm	0.085 km	85 m
		8.5m	85 000 mm
7	0.3 km	30 m	30 000 cm
		3.0 m	3000 cm

Convert the following.

8 2000 m = ______________ km

9 14 cm = ______________ mm

10 3.5 km = ______________ m

11 7 m = ______________ cm

12 6800 cm = ______________ m

13 60 mm = ______________ cm

14 5 cm = ______________ mm

15 800 m = ______________ km

16 350 cm = ______________ m

17 9 mm = ______________ cm

Highlight the greater length.

18 15 cm 160 mm

19 450 cm 4.6 m

20 18 000 m 1.9 km

21 203 mm 23 cm

22 0.2 m 21 mm

23 1050 m 1.5 km

24 5 mm 0.04 m

25 1.01 m 1001 mm

26 11 m 11 000 cm

27 3.6 km 360 000 mm

N

Mass

- The basic unit for measuring mass is the **gram**.
- All other units of mass in the metric system are based on the gram.
- Mass is often mistakenly called weight.

Weight is a measure of the pull of gravity on an object and is measured in **Newtons**.

Mass is the amount of matter an object contains and is measured in **grams**.

Use the following chart to help you convert mass.

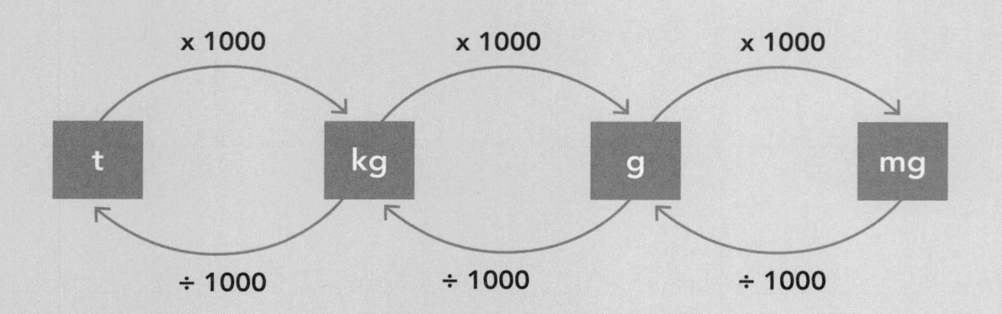

There are 1000 kg in 1 t.

Example:

1 How many tonnes are there in 3600 kilograms? 3600 kg = (3600 ÷ 1000 t)
= 3.6 t

2 How many milligrams are there in 5.6 grams? 5.6 g = 5.6 x 1000 mg
= 5600 mg

There are 1000 mg in 1 g.

28 Tick the correct calculation for these conversions.

a 3.4 kilograms into grams

☐ 3.4 x 1000 ☐ 3.4 x 100 ☐ 3.4 ÷ 1000 ☐ 3.4 ÷ 100

b 27 000 grams into tonnes

☐ 27 000 ÷ 1000 x 1000 ☐ 27 000 x 1000 ÷ 1000

☐ 27 000 x 1000 x 1000 ☐ 27 000 ÷ 1000 ÷ 1000

c 0.8 kilograms into milligrams

☐ 0.8 ÷ 1000 x 1000 ☐ 0.8 x 1000 ÷ 1000

☐ 0.8 x 1000 x 1000 ☐ 0.8 ÷ 1000 ÷ 1000

ISBN: 9780170477710

Highlight the correct conversion for each of the following.

29	2 kg	2000 mg	0.2 t
		0.02 t	2000 g
31	3 t	30 000 kg	3000 g
		3000 kg	300 000 g
33	15 000 g	150 kg	51 kg
		15 kg	1500 kg

30	5000 g	50 kg	0.5 kg
		5 kg	50 000 mg
32	45 000 mg	450 g	0.45 kg
		45 g	4.5 kg
34	0.7 t	7000 kg	700 kg
		70 000 g	70 000 kg

Convert the following.

35 38 kg = ____________ g

36 8 t = ____________ kg

37 76 000 mg = ____________ g

38 7900 g = ____________ kg

39 1.5 t = ____________ kg

40 840 kg = ____________ t

41 0.6 g = ____________ mg

42 390 g = ____________ kg

43 29 kg = ____________ t

44 935 mg = ____________ g

Highlight the larger mass.

45 60 g   6 000 mg

46 8.9 kg   89 000 g

47 3000 mg   30 g

48 0.7 t   70 kg

49 30 kg   0.003 t

50 850 000 mg   8500 g

51 0.4 kg   40 000 mg

52 0.2 t   2 million g

53 1 420 000 g   14.2 t

54 6 million mg   0.06 t

N

Capacity (volume)

- The basic unit for measuring capacity is the **litre**.
- All other units of capacity in the metric system are based on the litre.

Use the following chart to help you convert capacity.

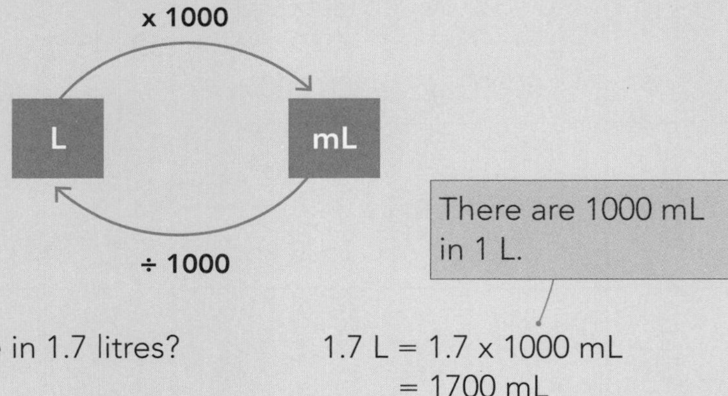

Example: How many millilitres are in 1.7 litres?

1.7 L = 1.7 x 1000 mL
= 1700 mL

Highlight the correct conversion for each of the following.

55	9000 mL	90 L	9 L
		0.09 L	0.9 L
57	0.08 L	80 mL	800 mL
		8000 mL	8 mL

56	1.5 L	150 mL	1550 mL
		1500 mL	15 mL
58	24 000 mL	2.4 L	240 L
		24 L	0.24 L

Convert the following.

59 9 L = ______________ mL

60 7300 mL = ______________ L

61 42 000 mL = ______________ L

62 5.6 L = ______________ mL

63 29 L = ______________ mL

64 550 mL = ______________ L

65 20 mL = ______________ L

66 0.2 L = ______________ mL

ISBN: 9780170477710

Perimeter

- Remember, to find the perimeter you need **start at one corner** and **add** the distances around the outside of a shape.
- None of these shapes is drawn to scale.

All the sides are the same, because this is a square.

These symbols tell you that these two sides are equal.

Examples:

1

9 m

9 m

Perimeter = 9 + 9 + 9 + 9
= 36 m

2

10 cm

6 cm

Perimeter = 6 + 10 + 10
= 26 cm

Calculate the perimeters of these shapes.

1

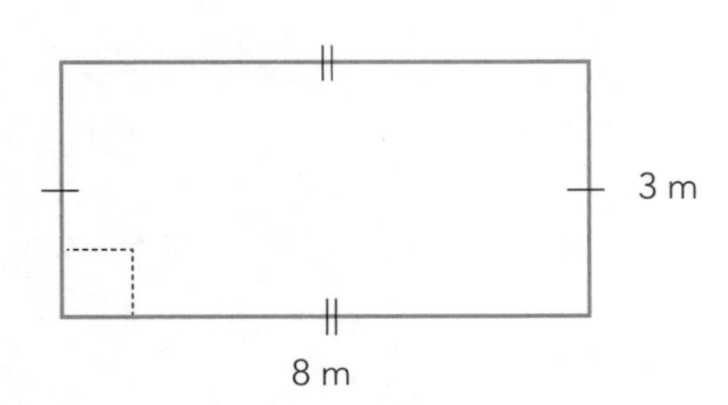

____

____

2

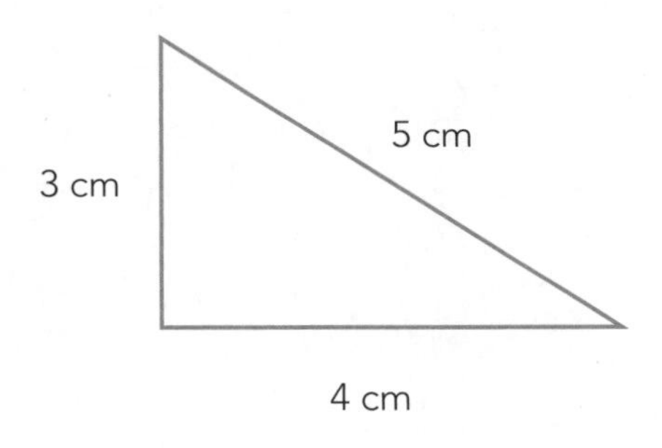

____

____

3

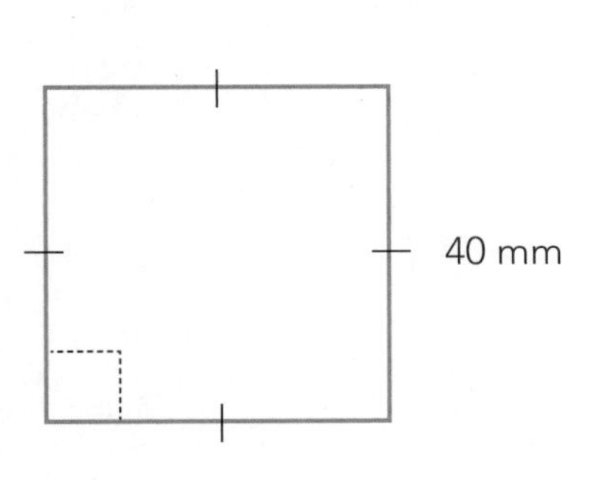

____

____

4

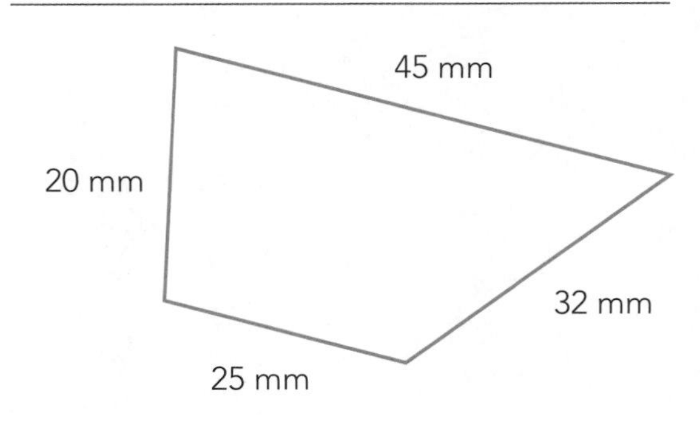

____

____

ISBN: 9780170477710

N

5

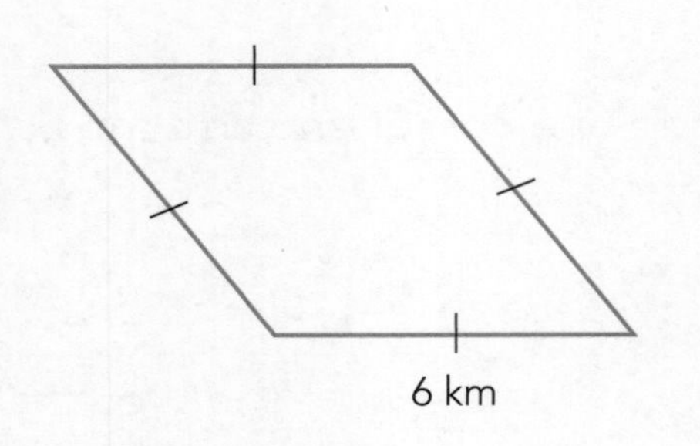

6

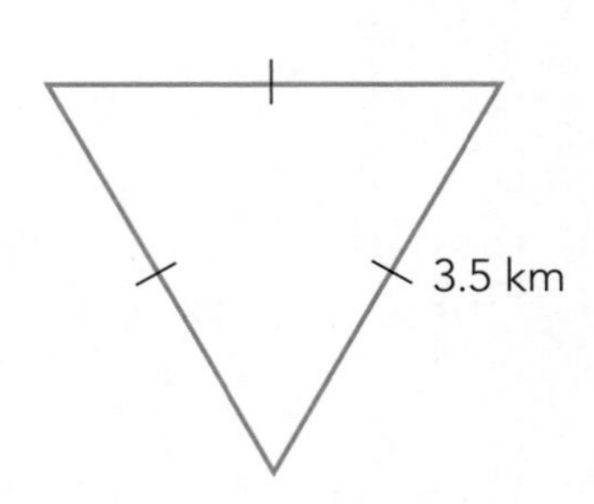

7

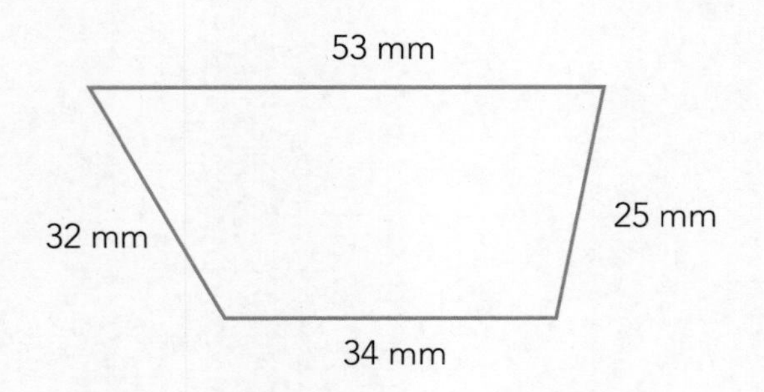

8

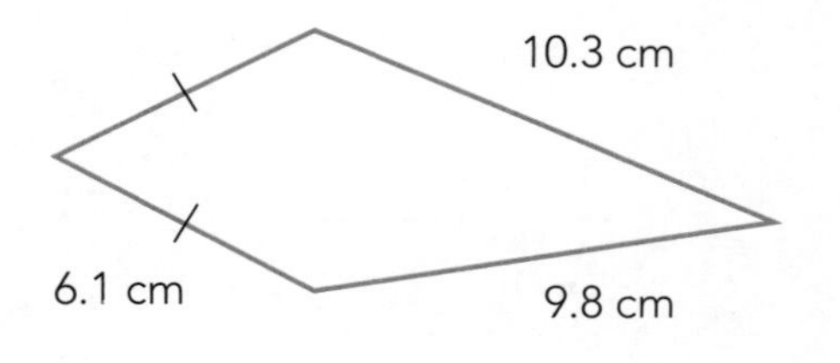

9

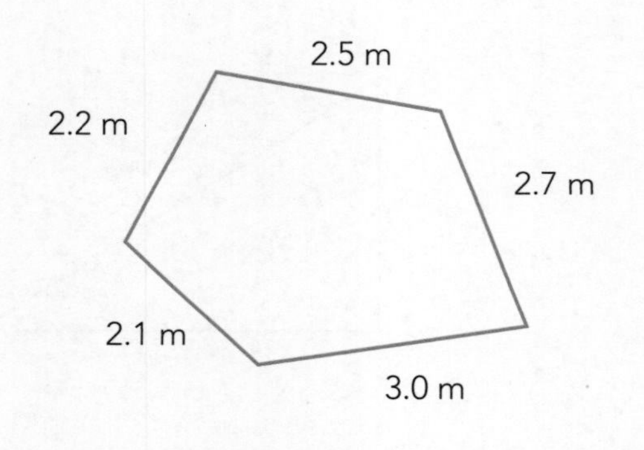

10

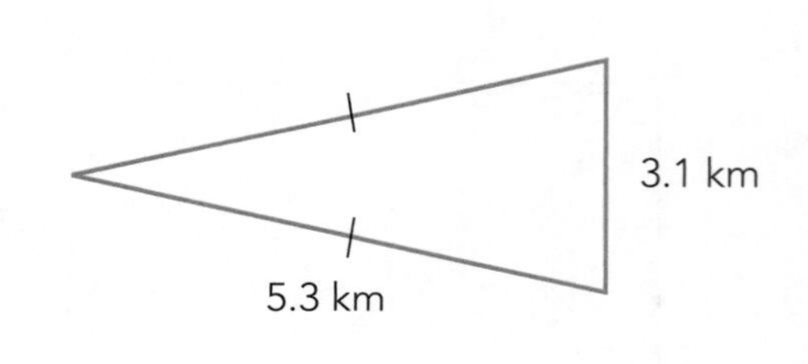

11

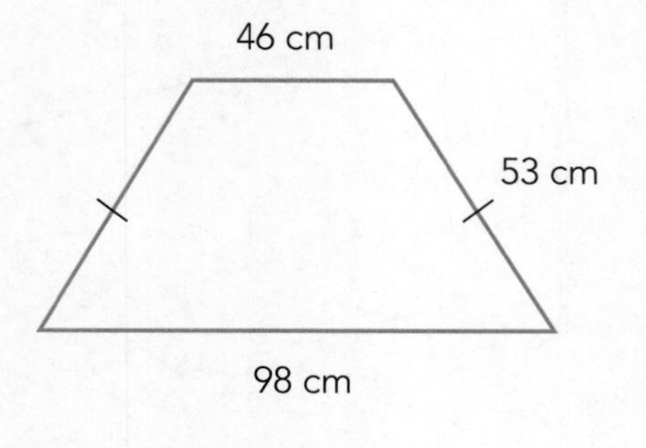

12

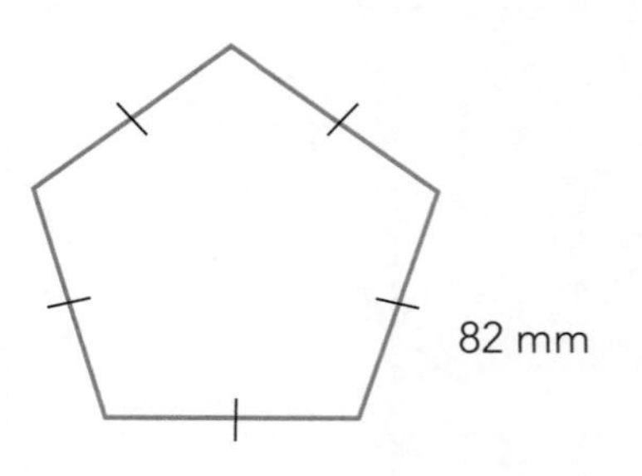

ISBN: 9780170477710

Things to look out for

Regular shapes

- 'Regular' means all the sides are the **same length**. In this case, each side is 11 cm long.

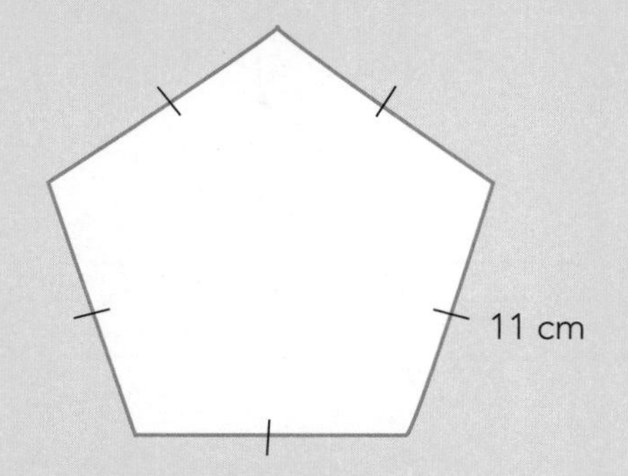

Perimeter = 11 + 11 + 11 + 11 + 11

= 5 x 11

= 55 cm

Extra measurements

- There may be measurements that are **not needed** for calculating the perimeter.

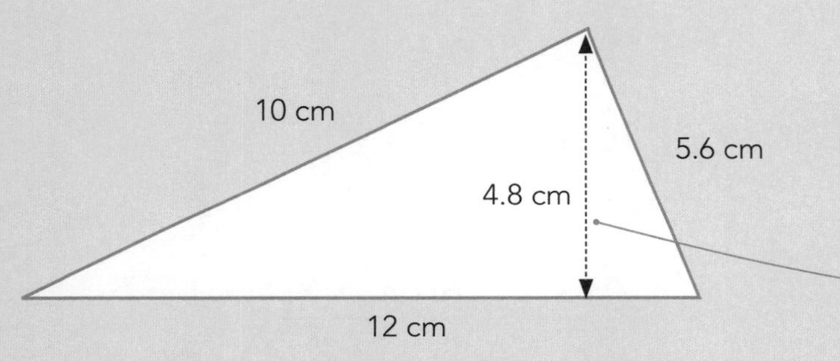

Perimeter = 10 + 5.6 + 12

= 27.6 cm

The 4.8 cm is not needed in order to calculate the perimeter.

Different units

- Some shapes may have measurements with **different units**.
- You need to make sure that all the information you need is in the same units before calculating the perimeter.

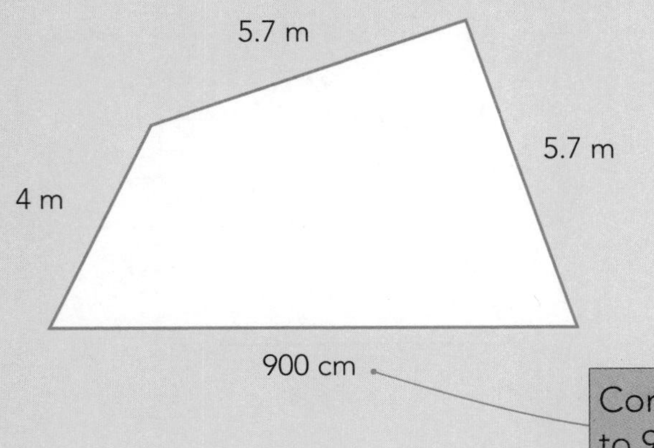

Perimeter = 4 + 5.7 + 5.7 + 9

= 24.4 m

Convert 900 cm to 9 m.

ISBN: 9780170477710

N

Calculate the perimeters of these shapes.

13

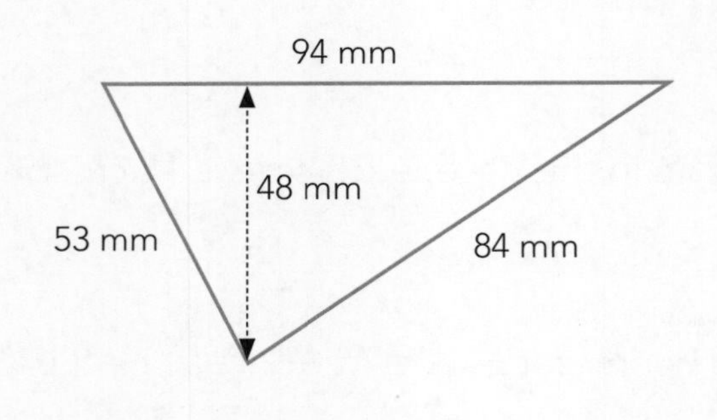

14 This is a regular hexagon.

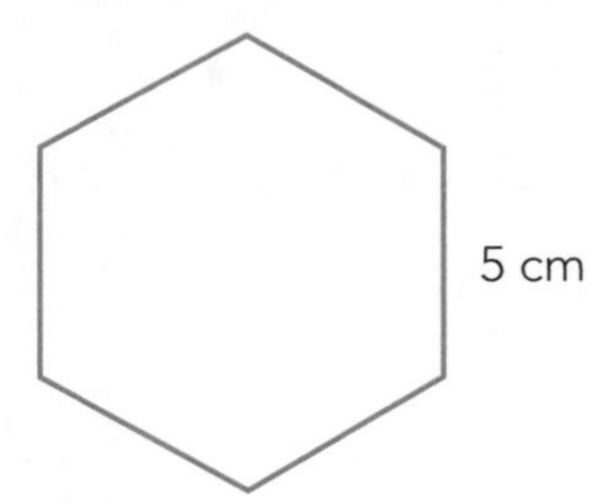

15

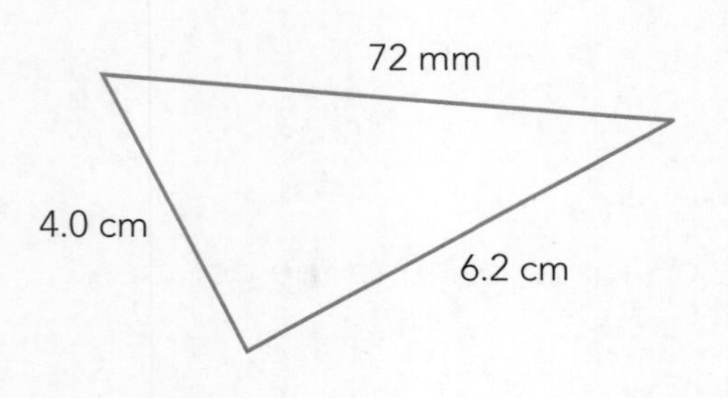

16

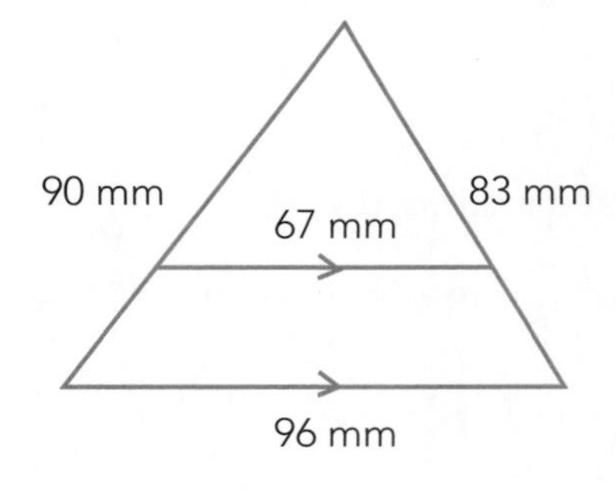

17 This is a regular octagon.

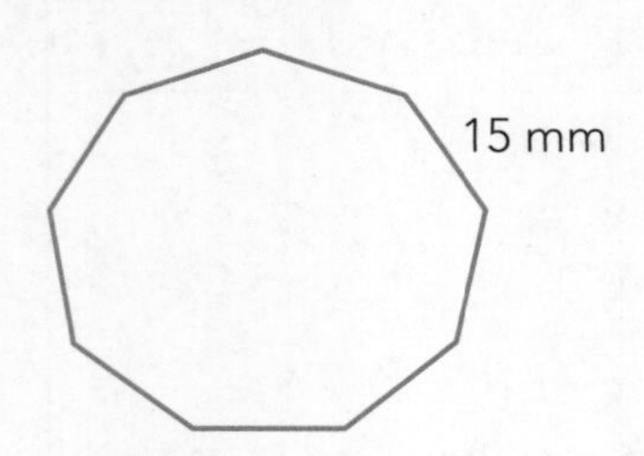

18

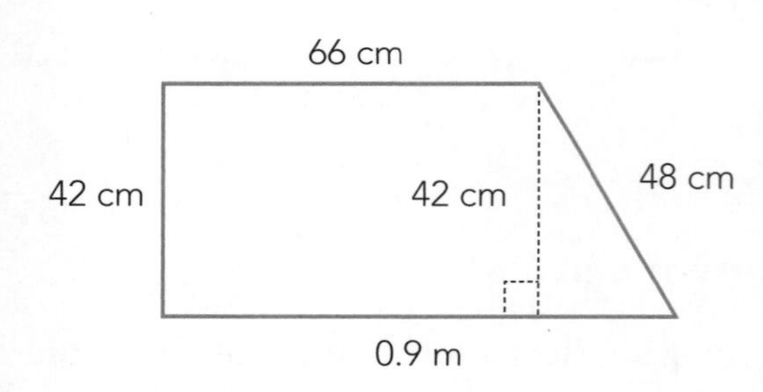

19

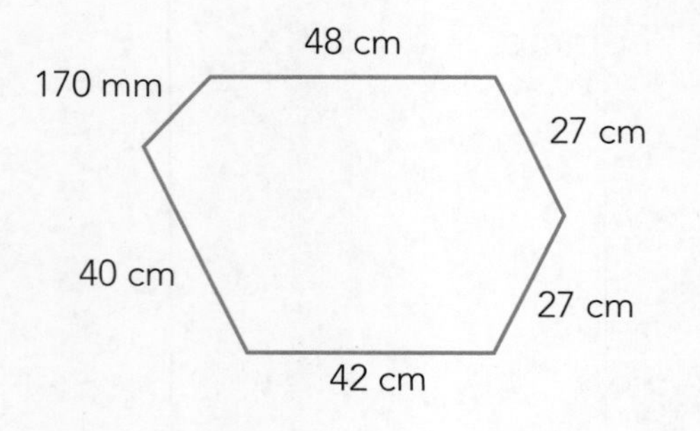

20

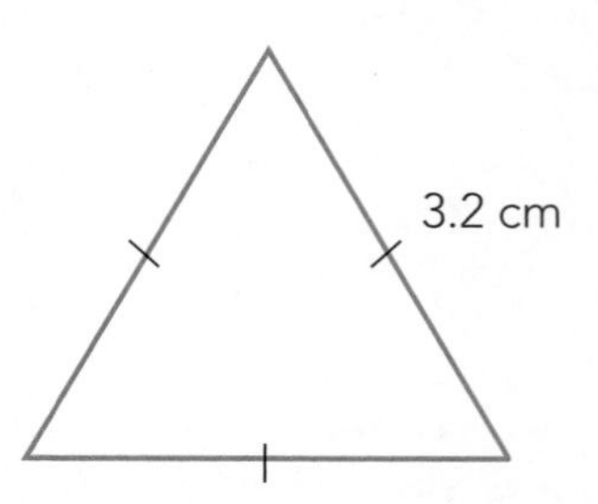

 ISBN: 9780170477710

Area

- Areas are measured in **square units**, e.g. m^2, cm^2, mm^2, km^2.
- One hectare is 10 000 m^2, and equivalent to a 100 m x 100 m square.

Squares and rectangles

- The area of a square or rectangle is calculated by multiplying the base by the vertical height.

Area = base x vertical height

$A = b \times h$

Examples:

1

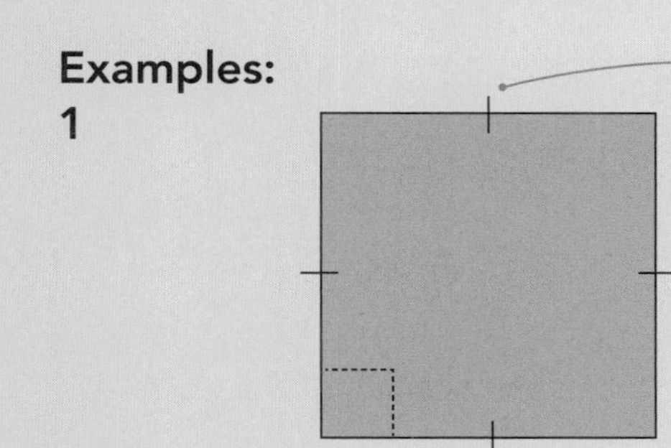

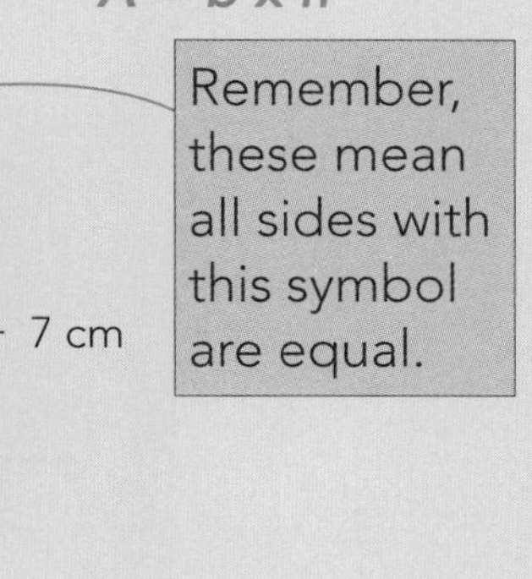

Area = $b \times h$
$= 7 \times 7$
$= 49\ m^2$

Units for area **must** be squared (^{2}).

2

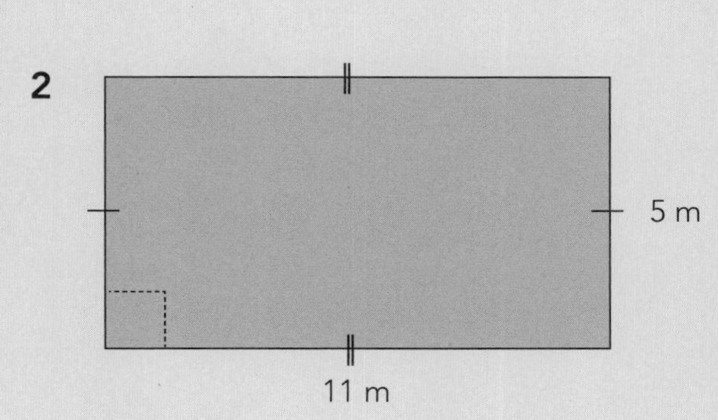

Area = $b \times h$
$= 11 \times 5$
$= 55\ m^2$

Calculate the areas of these shapes.

1

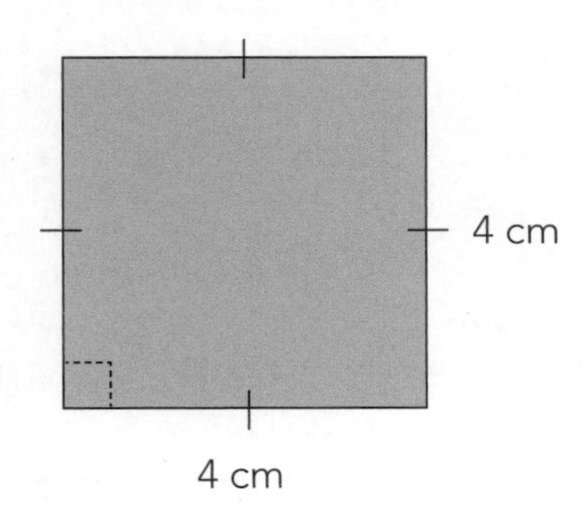

____

____

2

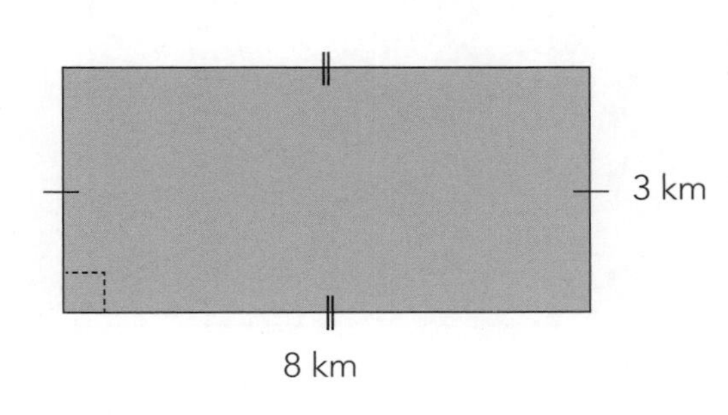

____

____

3

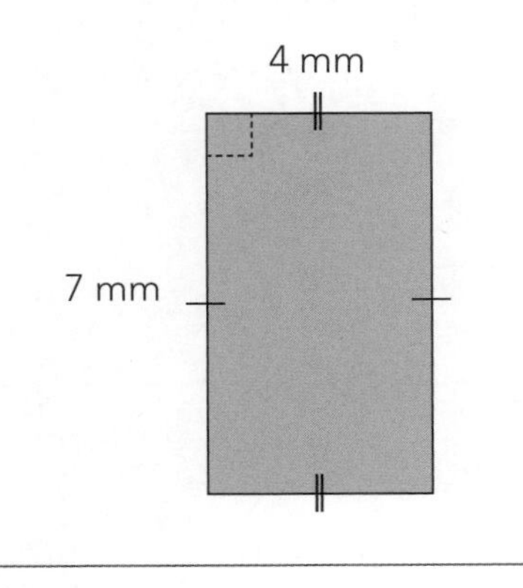

____

____

4

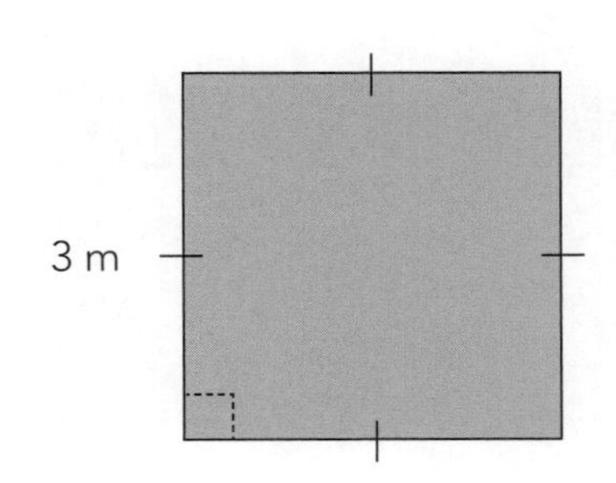

____

____

ISBN: 9780170477710

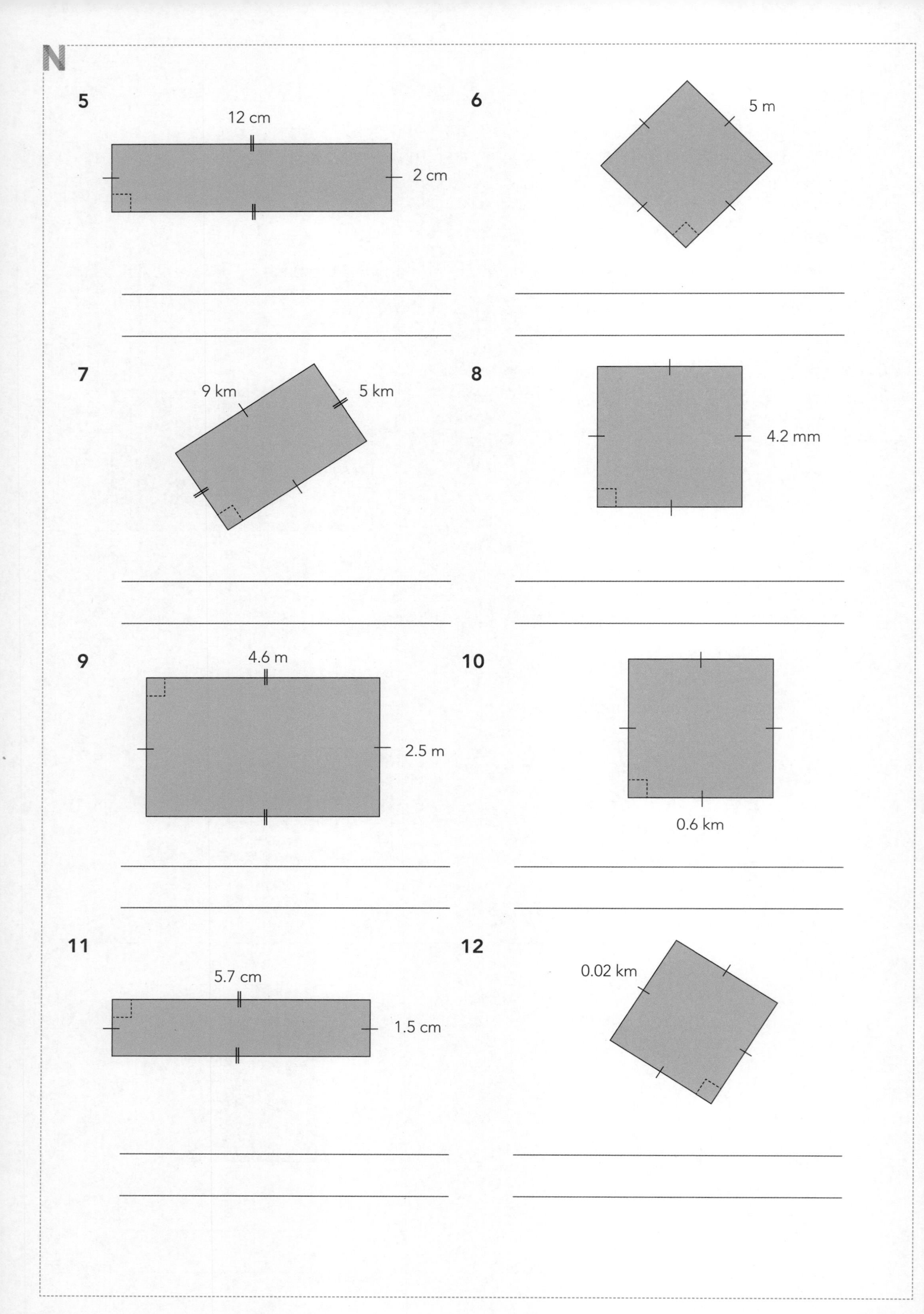

ISBN: 9780170477710

Parallelograms and rhombuses

- A **parallelogram** has opposite sides that are parallel and equal.
- A **rhombus** has four equal sides, and opposite sides are parallel.
- For square units the number you use must be at right angles.

You can rearrange a parallelogram to look like a rectangle.

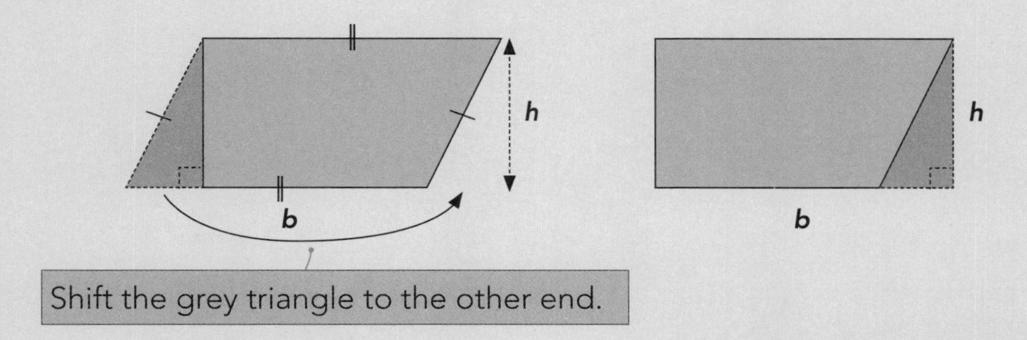

So the formula stays the same: **Area = base x vertical height**

$$A = b \times h$$

Examples:

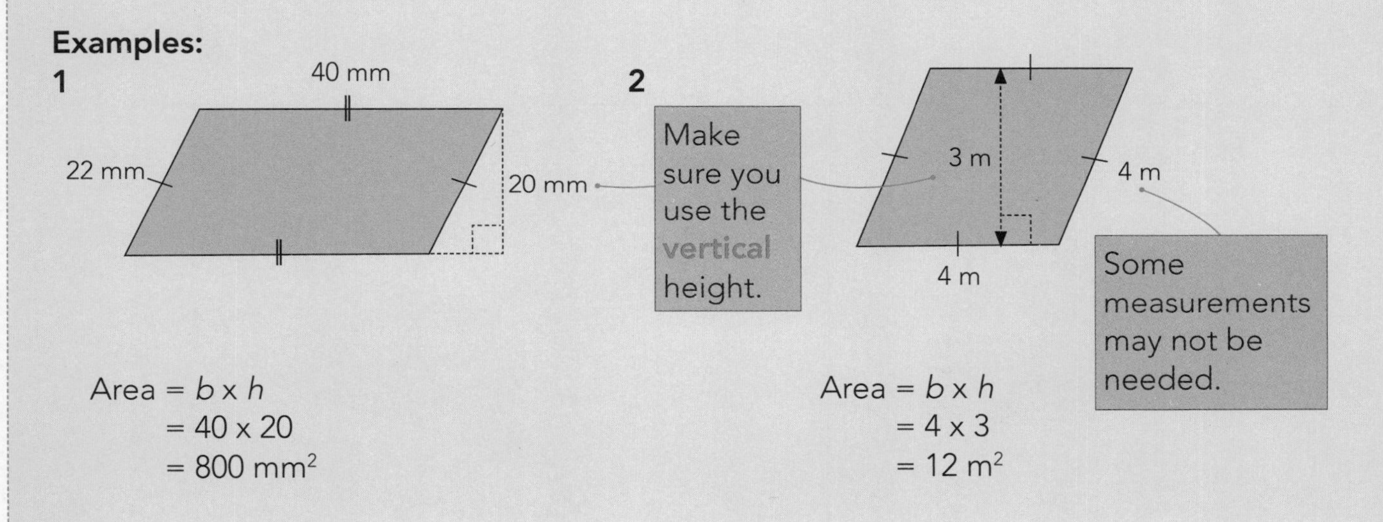

Area = $b \times h$
= 40×20
= 800 mm^2

Area = $b \times h$
= 4×3
= 12 m^2

Calculate the areas of these shapes.

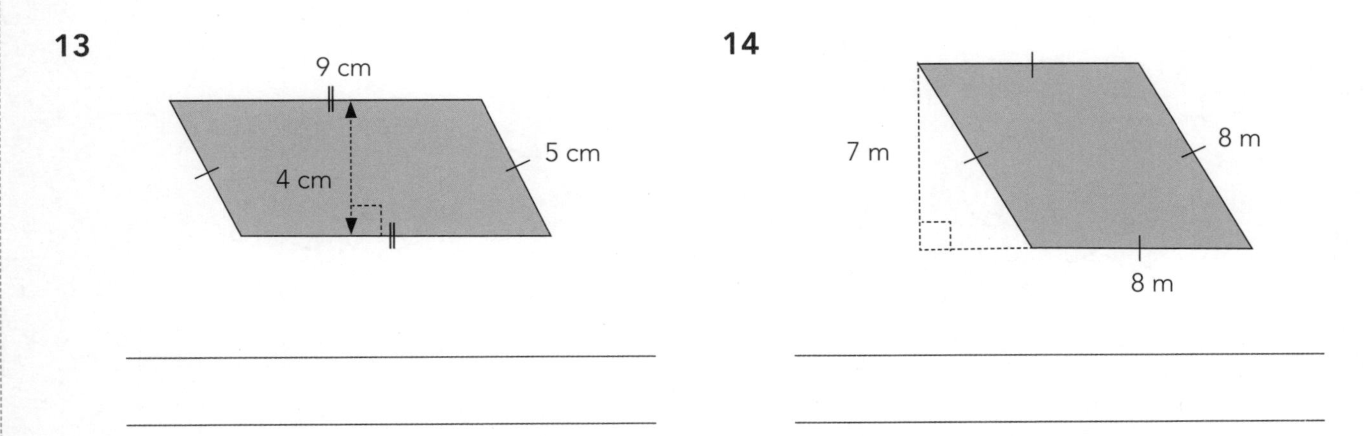

ISBN: 9780170477710

N

15

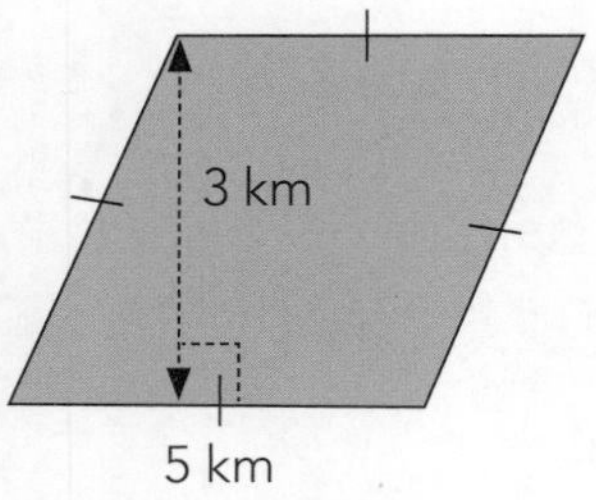

3 km

5 km

16

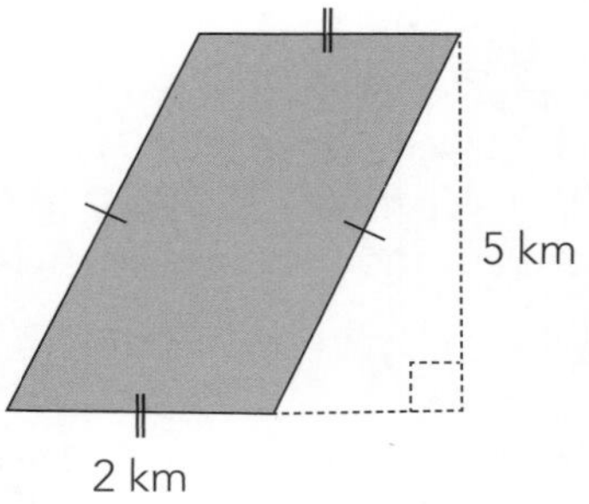

5 km

2 km

17

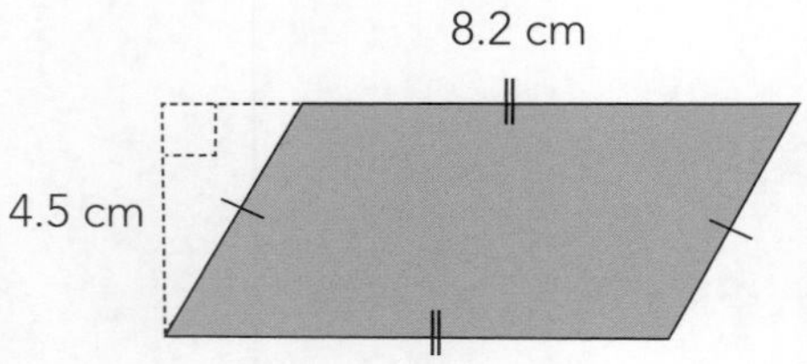

8.2 cm

4.5 cm

18

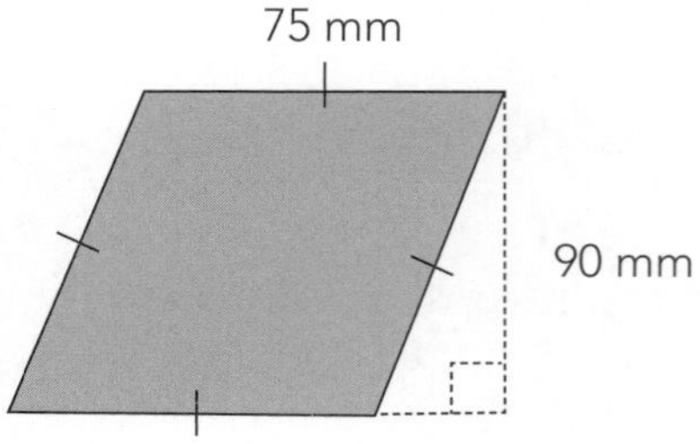

75 mm

90 mm

19

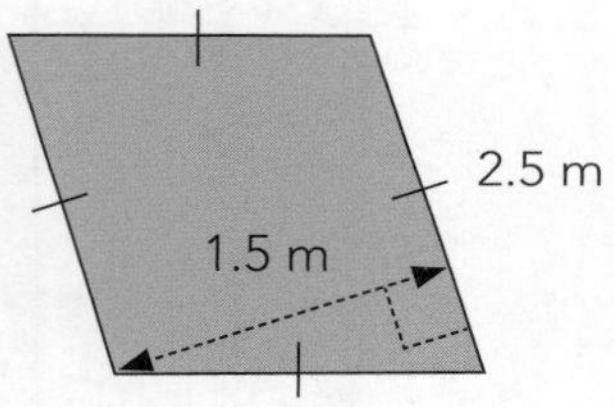

2.5 m

1.5 m

20

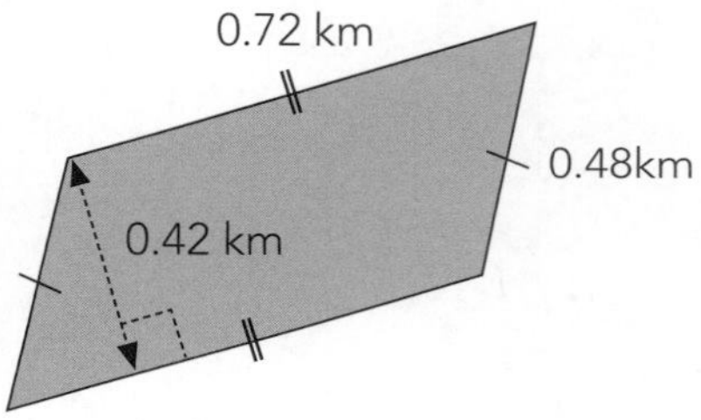

0.72 km

0.48km

0.42 km

21

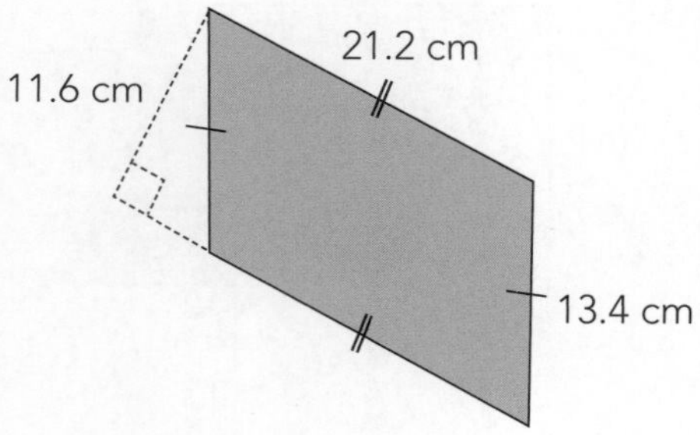

21.2 cm

11.6 cm

13.4 cm

22

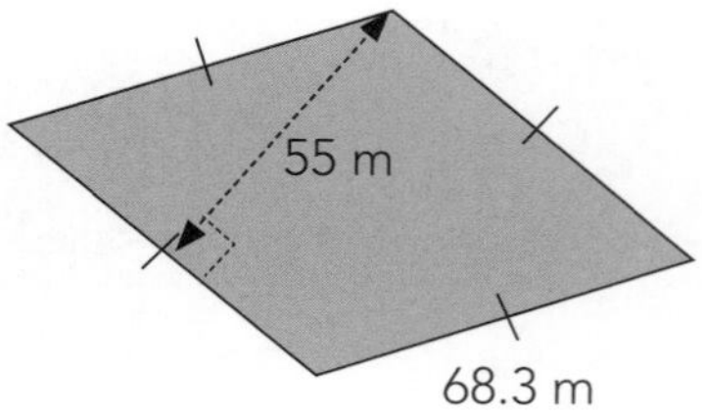

55 m

68.3 m

 ISBN: 9780170477710

Triangles

- A **triangle** is a closed figure with three straight sides and three angles.
- The area is half its base multiplied by the vertical height.

$$\text{Area} = \frac{1}{2} \times \text{base} \times \text{vertical height}$$

$$A = \frac{1}{2} \times b \times h$$

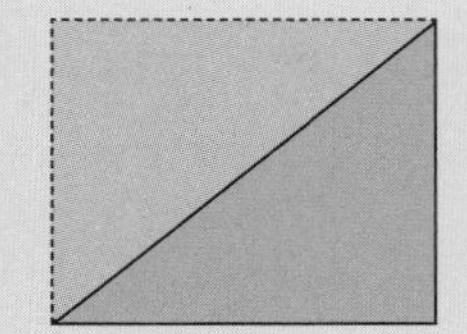

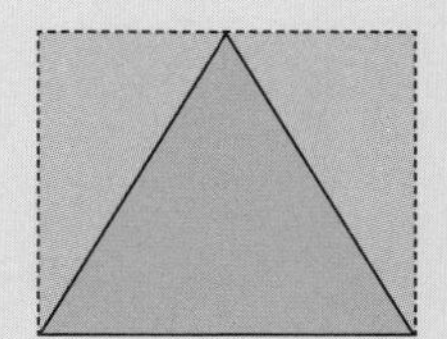

$$\text{Area} = \frac{1}{2} \times (\text{area rectangle}) = \frac{1}{2} \times \text{base} \times \text{height} = \frac{1}{2} \times b \times h$$

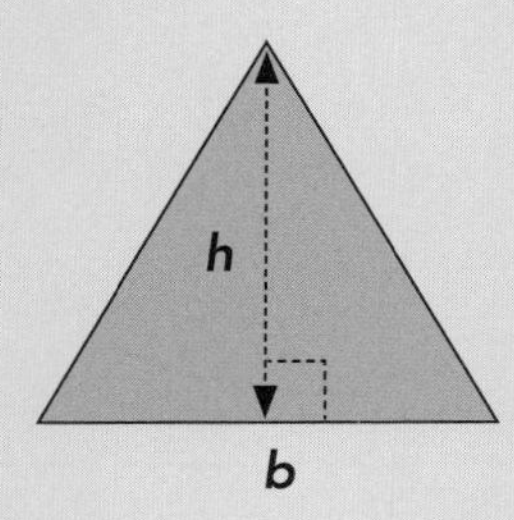

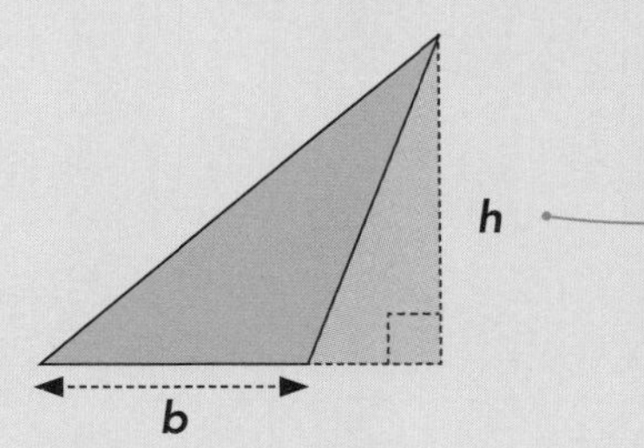

You must use the **vertical** height.

Examples:

1

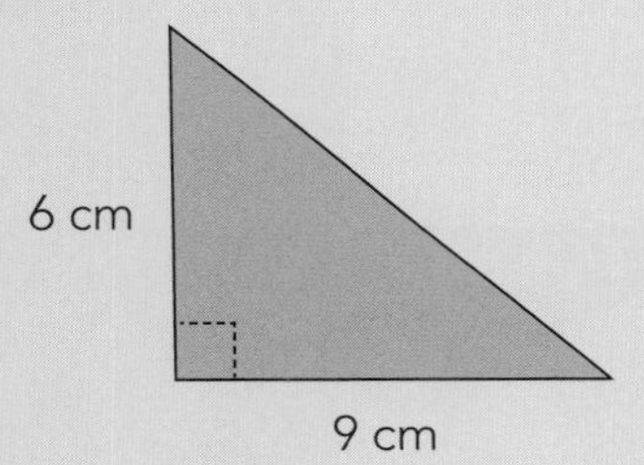

$\text{Area} = \frac{1}{2} \times b \times h$

$= \frac{1}{2} \times 9 \times 6$

$= 27 \text{ cm}^2$

2

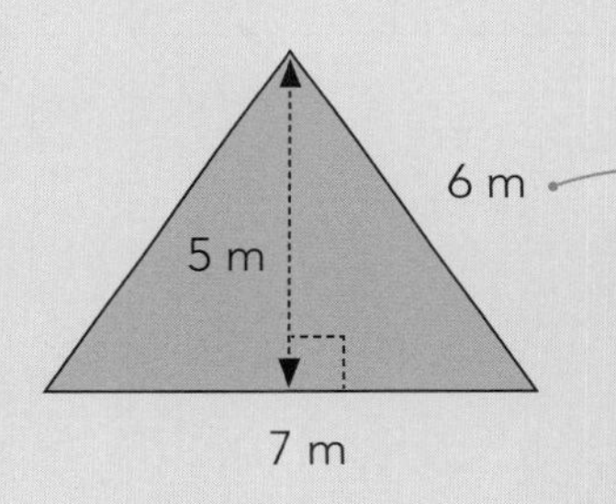

Some measurements may not be needed.

$\text{Area} = \frac{1}{2} \times b \times h$

$= \frac{1}{2} \times 7 \times 5$

$= 17.5 \text{ m}^2$

Calculate the areas of these shapes.

23

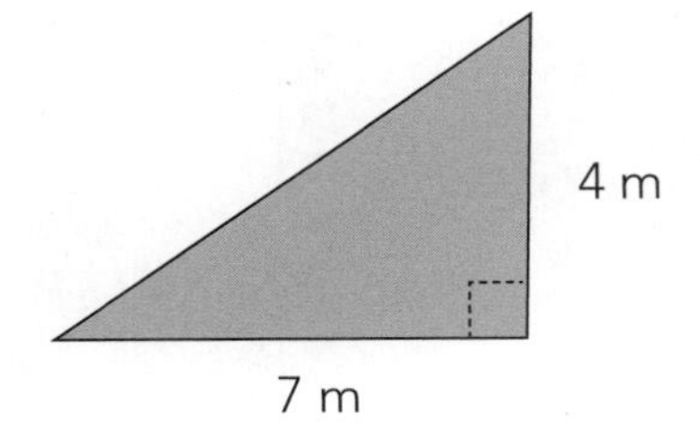

____

____

24

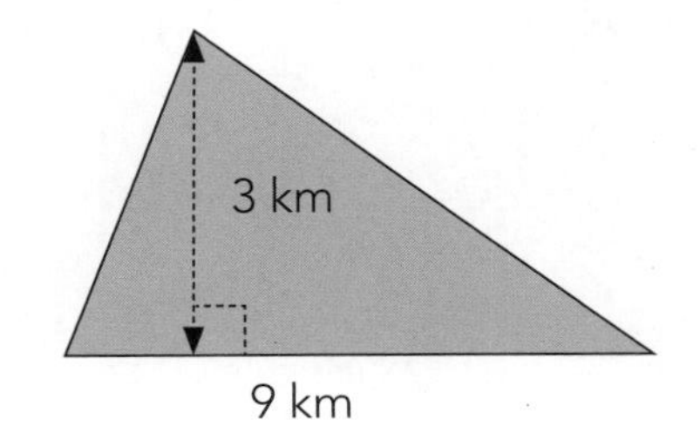

____

____

ISBN: 9780170477710

N

25

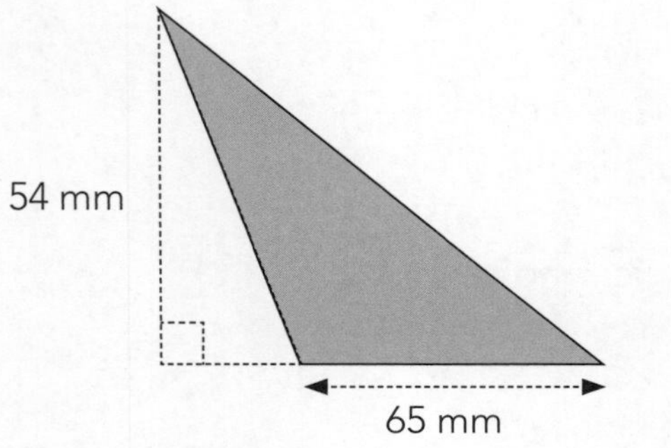

54 mm
65 mm

26

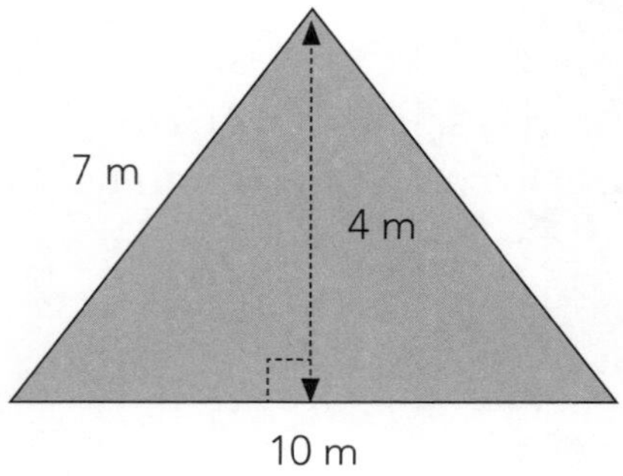

7 m
4 m
10 m

27

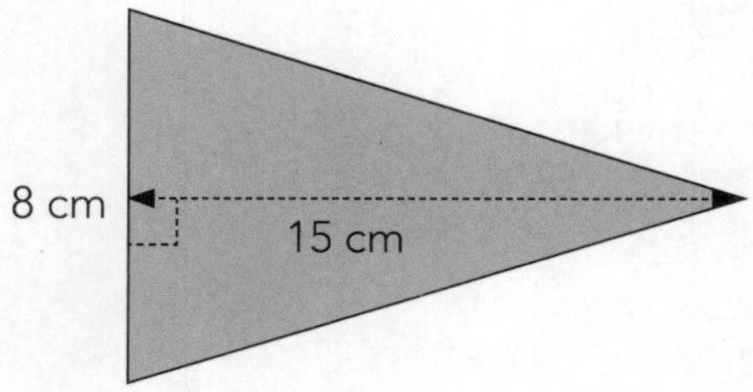

8 cm
15 cm

28

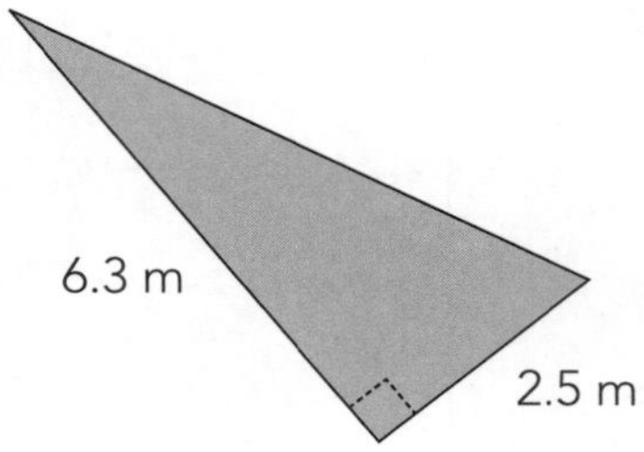

6.3 m
2.5 m

29

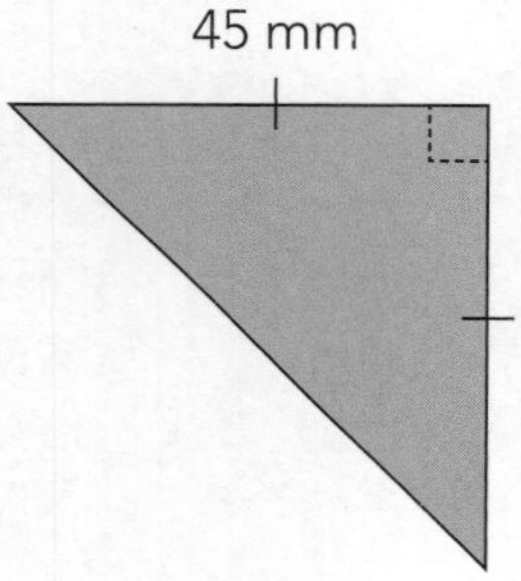

45 mm

30

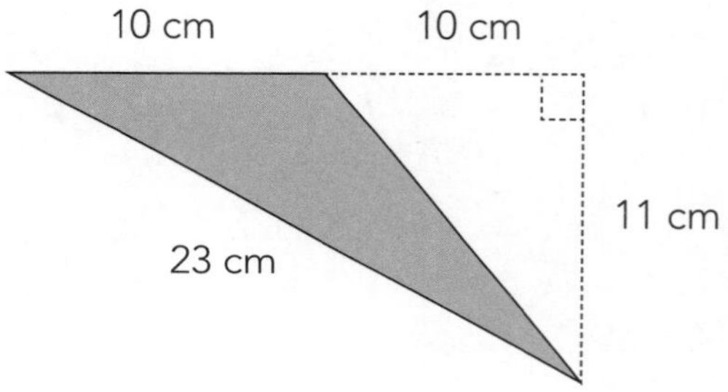

10 cm
10 cm
11 cm
23 cm

31

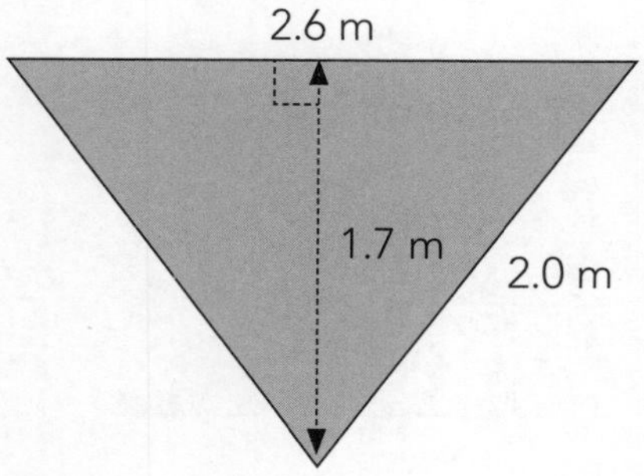

2.6 m
1.7 m
2.0 m

32

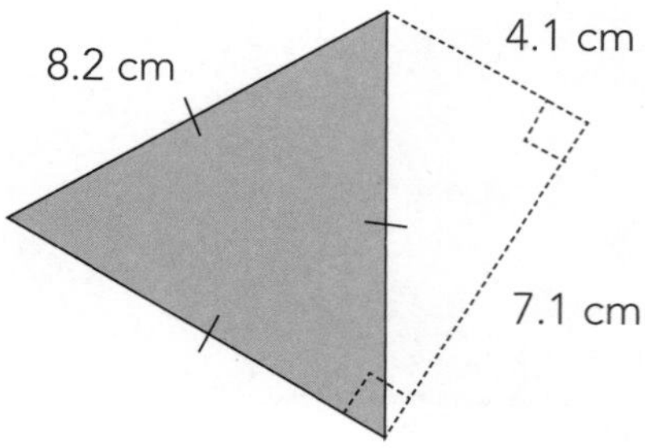

8.2 cm
4.1 cm
7.1 cm

 ISBN: 9780170477710

Volume

- Volume is measured in **cubic units**, e.g. m^3, cm^3, mm^3, km^3.

Cuboids with cube blocks

- The volume of a three-dimensional (3D) shape is the amount of **space** the shape occupies.
- A **cuboid** is a box shape, e.g. a shoebox.
- A **cube** is a box shape where all the dimensions are equal, e.g. a die.
- Each block represents 1 cm^3.

Volume = height x width x depth

$V = h \times w \times d$

Example:

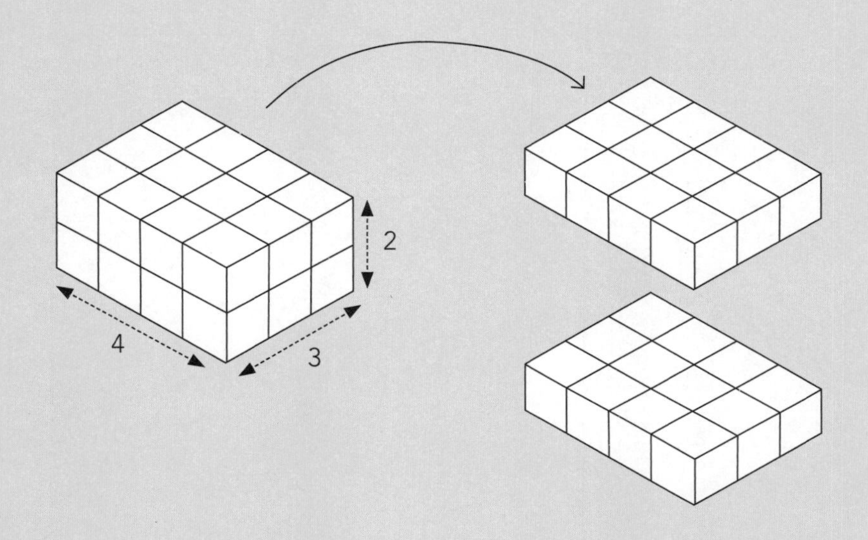

Volume = number of layers x number of blocks in one layer

= 2 x 4 x 3

= 24 cm^3

This is the same as **height x width x depth**.

When calculating volume, the units **must** be cubed (^{3}).

Calculate the volumes of these cuboids. Each block represents 1 cm^3.

1

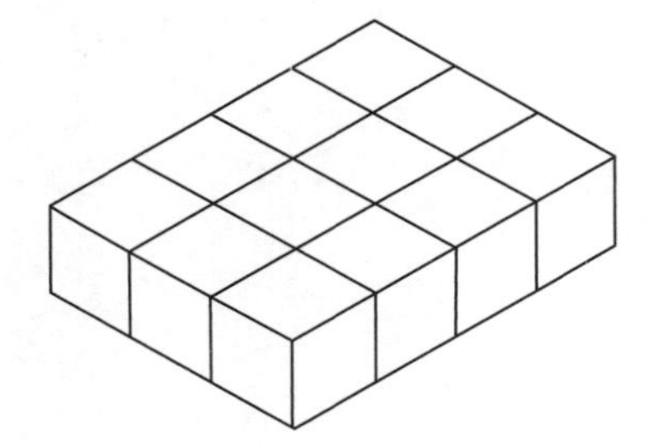

Volume = $h \times w \times d$

= ______________

2

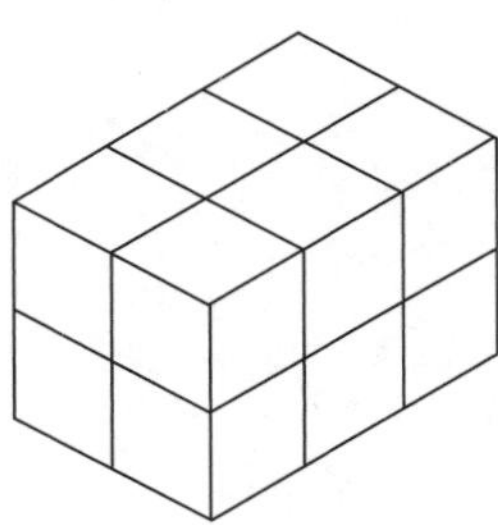

____

____

ISBN: 9780170477710

N

3

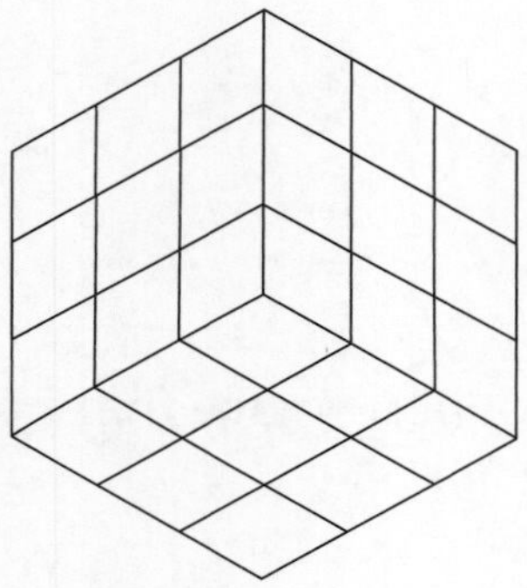

4

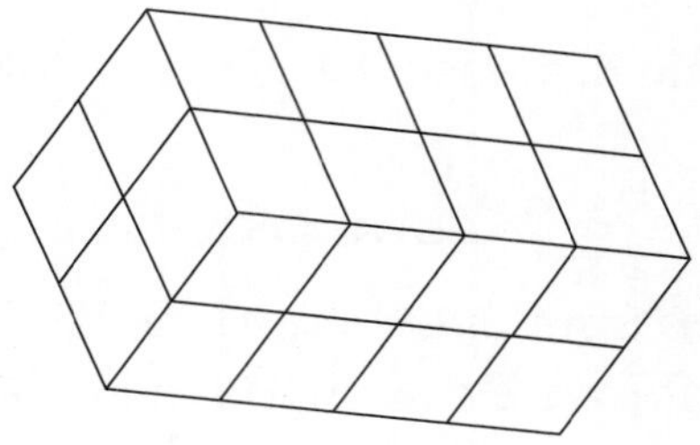

5

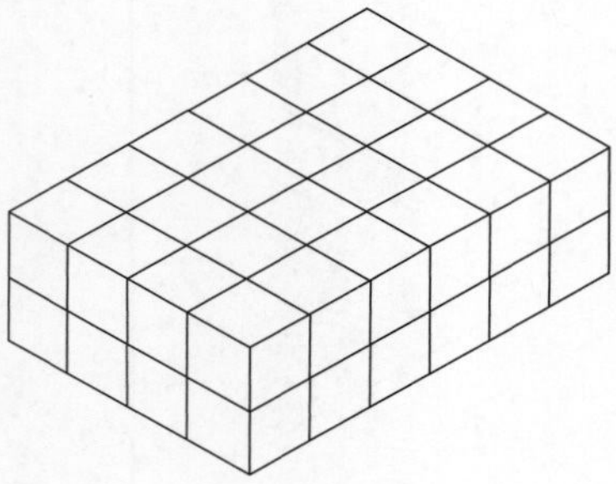

6

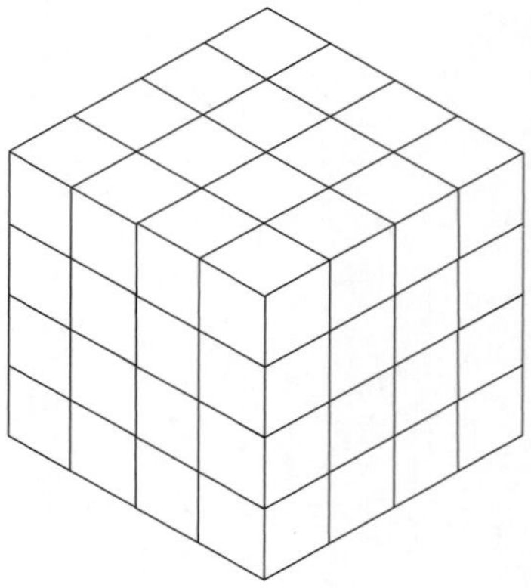

7

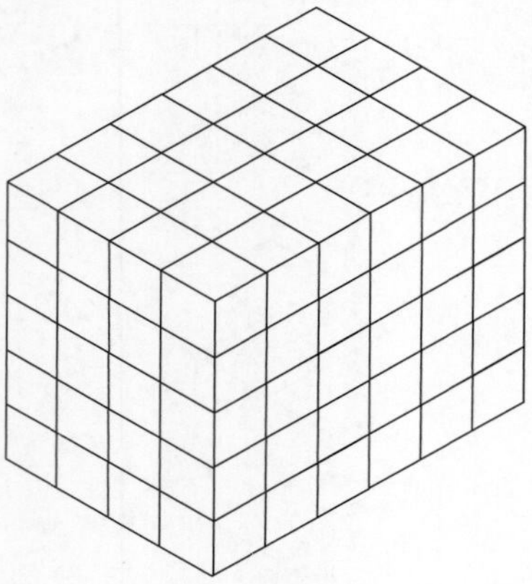

8

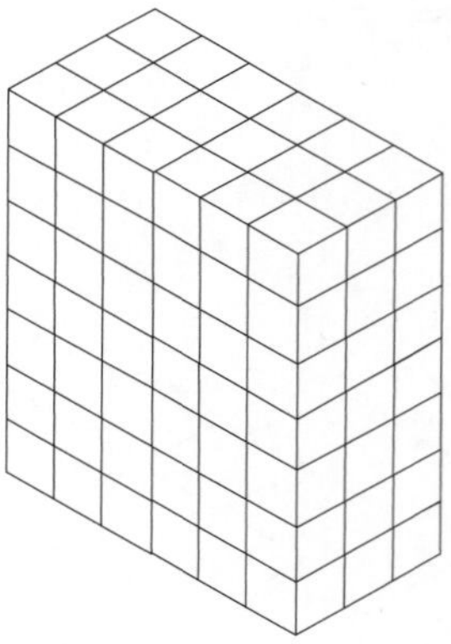

 ISBN: 9780170477710

Other cuboids

Volume = height x width x depth

$V = h \times w \times d$

Examples:

1

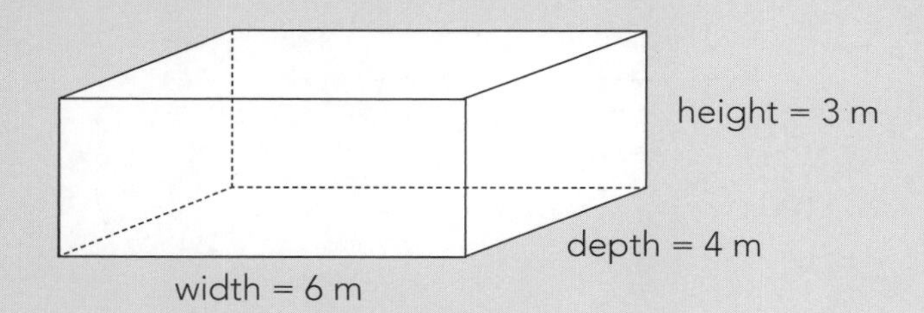

Volume = $h \times w \times d$

$= 3 \times 6 \times 4$

$= 72\text{ m}^3$

2 Another way to think about it is to find the area of the 'face' and multiply it by the depth.

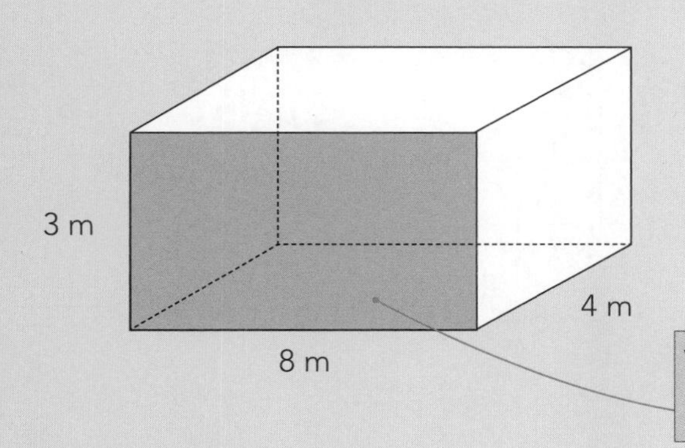

Volume = **area of face** x d

= **(3 x 8)** x 4

= **(24)** x 4

= 96 m^3

Volume = area of face (24 m^2) x depth (4 m)
= 96 m^3

Calculate the volumes of these cuboids.

9

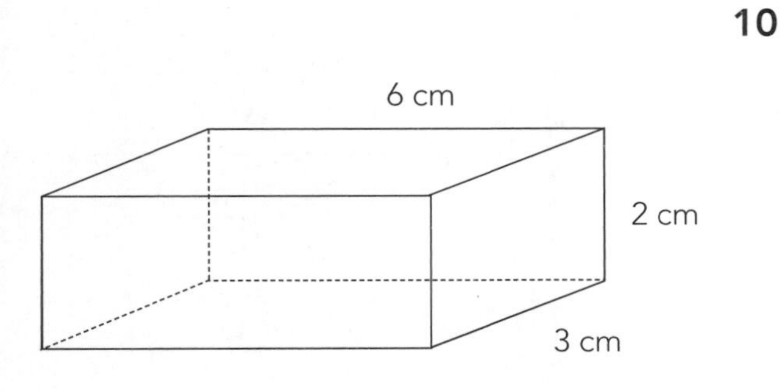

____

____

10

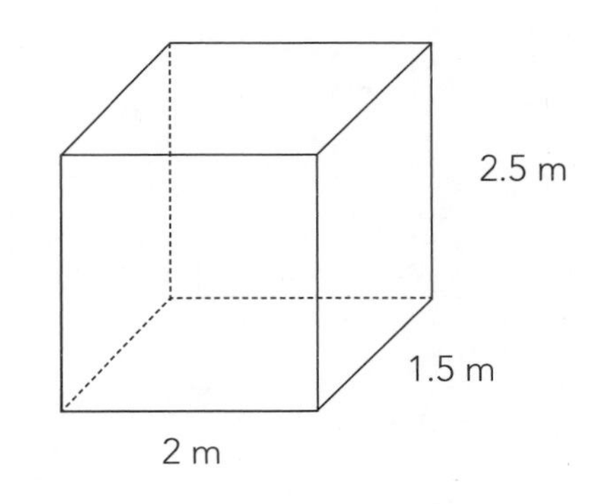

____

____

ISBN: 9780170477710

N

11

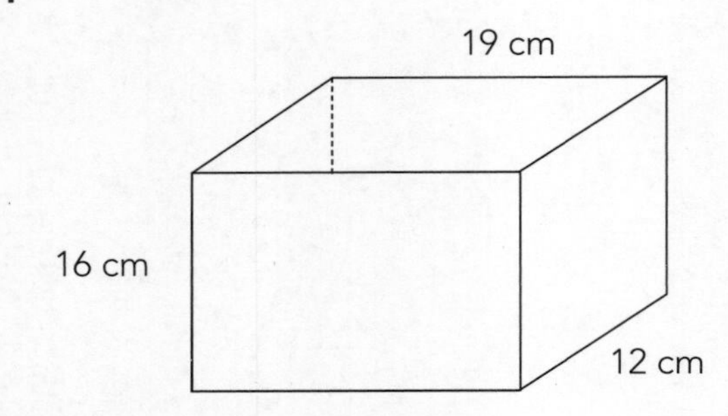

12

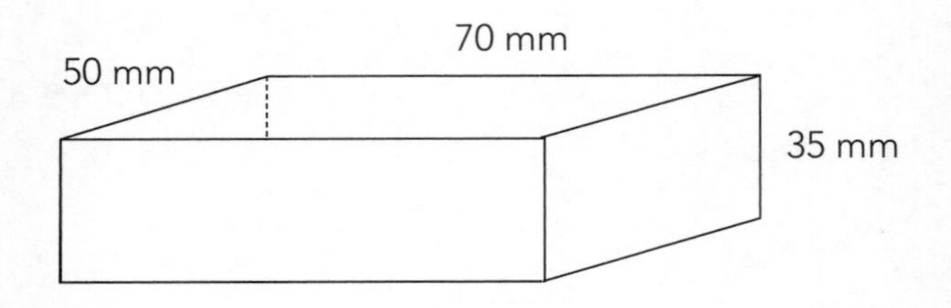

13

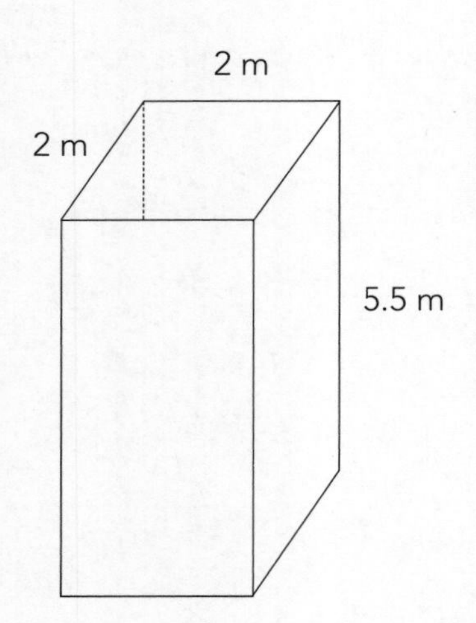

14 This is a cube.

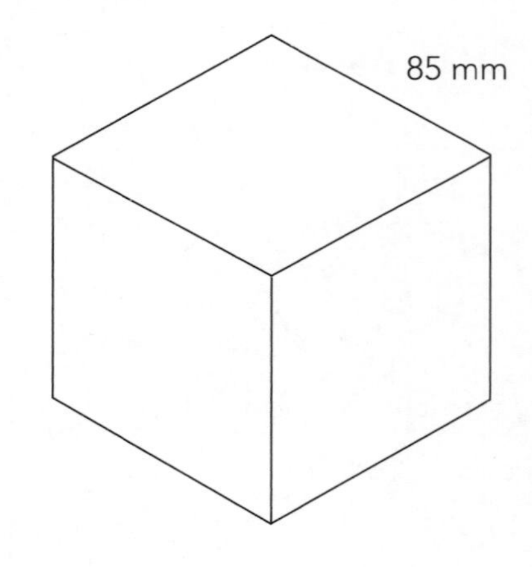

15

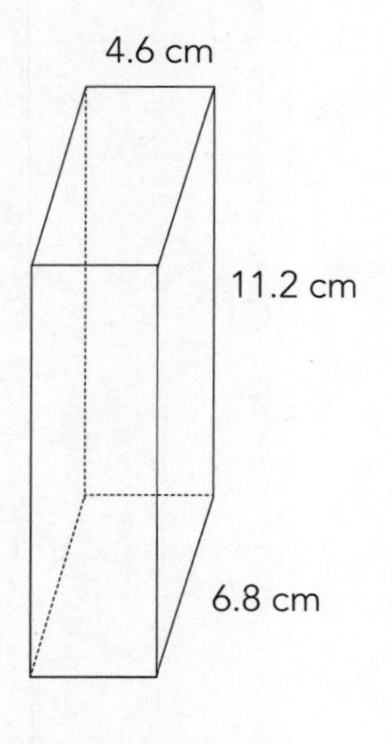

16

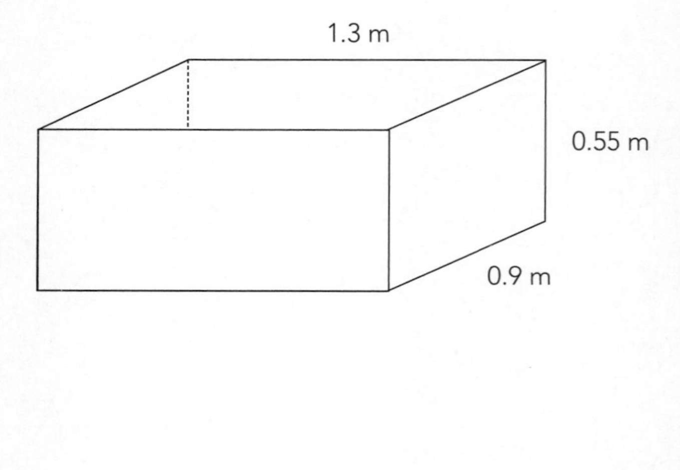

 ISBN: 9780170477710

Using rates

- Rates are often written using the word '**per**'.
- Rates are values with respect to another unit, often time.
- Examples: running speed in kilometres per hour;
  the rate at which a hot air balloon is inflated in cubic metres per minute;
  number of worms per hectare.

Example: Manaia is planning a day walk. The walk is 7.2 km and she estimates that she can walk at 4 km/hour. How long is this walk likely to take her?

This can also be written as 4 km/h.

Method 1

Step 1: Write the information you are given in a brief statement:

4 km takes **1 hour**

Because you are trying to find a time, put the **time** on the **right**.

Step 2: Underneath write a parallel statement about what you want to know:

Be careful when converting decimal hours to hours and minutes.
0.8 hours = 0.8 x 60 = 48 minutes.

4 km takes **1 hour**

copy

7.2 km takes **1** x $\frac{7.2}{4}$ = 1.8 hours = 1 hour and 48 minutes.

Multiply by either $\frac{7.2}{4}$ or $\frac{4}{7.2}$. Use $\frac{7.2}{4}$ here because 7.2 km will take **longer** than **4 km**.

Or

Method 2:

Step 1: Work out how long 1 km takes:

4 km takes **1 hour** or **60 min**

1 km takes $\frac{1}{4}$ **hour** or $\frac{60}{4}$ **min** = 0.25 hours or 15 minutes

Step 2: Multiply this time by the number of kilometres she will walk.

Walking 7.2 km will take 7.2 x 0.25 hours = 1.8 hours

= 1 hour and 48 minutes.

or Walking 7.2 km will take 7.2 x 15 minutes = 108 minutes

= 1 hour and 48 minutes.

N

Answer the following questions.

1 a Max lives 2.8 km from school. If he walks to school at 4 km/hour, how many minutes will it take him?

b He walks home from school at 3.5 km/hour, how many minutes will it take him?

2 Huia could feel her dog's heart beating while it sat on her lap. She managed to count 42 beats in 18 seconds before it jumped down. Calculate the dog's heart rate in beats per minute.

3 It took five builders 7.5 hours to build a fence. How long would it have taken if there were only three builders?

4 A flea can jump a distance of 200 times its body length. If Gia could jump the equivalent of a flea and she is 165 cm high, how far could she jump?

5 If 4 kg of oranges cost $23.96, how much would 3 kg of oranges cost?

6 A recipe requires 3 cups of flour for 15 cookies. How many cookies (of the same size) could be made with 7 cups of flour?

7 If 210 million snaps are created on Snapchat every day, how many are posted each hour?

ISBN: 9780170477710

8 Last year, Mere made 13 jars of jam from 3.25 kg of plums. This year she has 4.5 kg of plums to make into jam. How many jars can she expect to fill?

9 There is an average of 126 000 spiders per hectare (10 000 m^2) in green areas. How many spiders could you expect to have on a 7000 m^2 rugby field??

10 The Smith family go on holiday for two weeks leaving a tap dripping over the bath. The tap drips at a rate of one drip every two seconds, and 10 drips = 1 mL. How many litres of water will have been lost due to the dripping tap?

11 Ollie lives in a house where the water comes from two 23 000 L tanks. Before he left for holiday, both tanks were full. A hose fitting failed not long after he left, and it released water at 12 L per minute. To the nearest hour, after how long will the tanks be empty?

12 In 2022 a bar tailed godwit flew from Alaska to Tasmania in eleven days and one hour. During this time it did not drink, eat or land. The distance covered was 13 560 km. Calculate its average speed to the nearest km/hour.

13 Snails move at about 1 m per hour. At this rate, how long will it take a snail to move the 24 cm from one lettuce plant to the next? Give your answer in minutes and seconds.

14 Hair grows at 3.5 mm every ten days. To the nearest millimetre, how much does hair grow in a year (365 days)? Write your answer to the nearest millimetre.

ISBN: 9780170477710

N

Mixing it up

1 Brent is driving from Dunedin to Christchurch, a distance of 360 km.

a His average speed is 75 km/hour. If he has no breaks, how long will the drive take him? Give your answer in hours and minutes.

_______ hours _______ minutes

b On average, his car uses 8 L of fuel every 100 km. How much fuel will he use?

_______ L

c He stopped for lunch, so the total trip took him 5 hours and 32 minutes. He left Dunedin at 9:45 am. At what time did he reach Christchurch?

_______ am/pm

2 Huia is making a batch of scones.

a She needs 270 g of flour. Draw a straight line from the centre of the scale to through the 270 g mark on the scale.

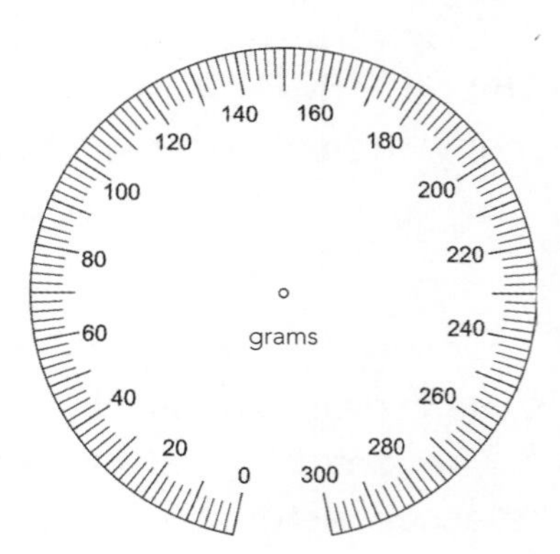

b She has a 2 kg bag of flour. How many batches of scones of this size could she make with 2 kg of flour?

_______ batches

c The recipe makes eight scones. If she wanted to make 12 scones, how much flour would she need?

_______ g

d Her scones need to cook for 25 minutes. If she wants to take them out of the oven at 3:15 pm, at what time should she put them in?

_______ pm

3 Isla is making a pen for her ducks.

a Calculate the perimeter of the pen.

Perimeter = ___________ m

3.9 m
2.6 m
3.3 m
Pen
3.5 m
2.1 m

b The netting she needs to put around the pen comes in 5 m rolls. How many rolls will she need to buy?

_______ rolls

c Hen eggs have a volume of about 60 mL. Her duck eggs have a volume of 75 mL. The recipe for the cake she is making needs five hen eggs. If she replaces these with duck eggs, how many should she use?

_______ duck eggs

ISBN: 9780170477710

4 This is part of the timetable for buses between Tūī Drive and Kiwi Road.

Tūī Drive	10:09	10:39	11:09	11:39
Pūkeko Lane	10:17	10:47	11:17	11:47
Korimako Ave	10:26	10:56	11:26	11:56
Taranui Place	10:33	11:03	11:33	12:03
Tōrea Street	10:43	11:13	11:43	12:13
Kiwi Road	10:48	11:18	11:48	12:18

a How often do buses leave from Tūī Drive?

Every ____________ min

b Adam catches the 11:47 bus from Pūkeko Lane. How long will it take him to get to Kiwi Road?

____________ min

c Eliza needs to be at Tōrea Street 11:45. What is the latest bus that she can catch from Pūkeko Lane?

____

5 Eric is making a sandpit. In total it will be 1.6 m long and 1.5 m wide.

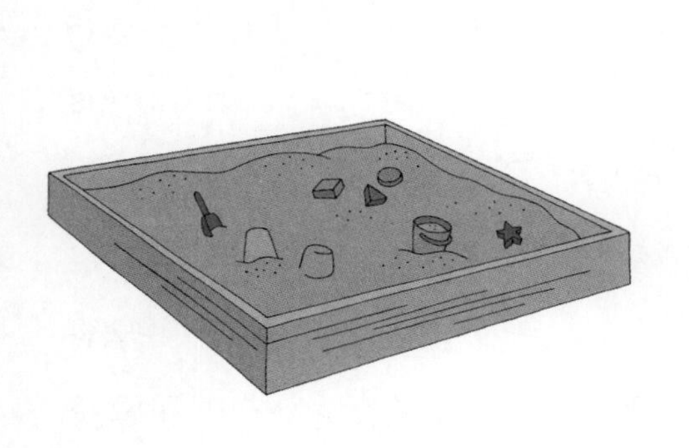

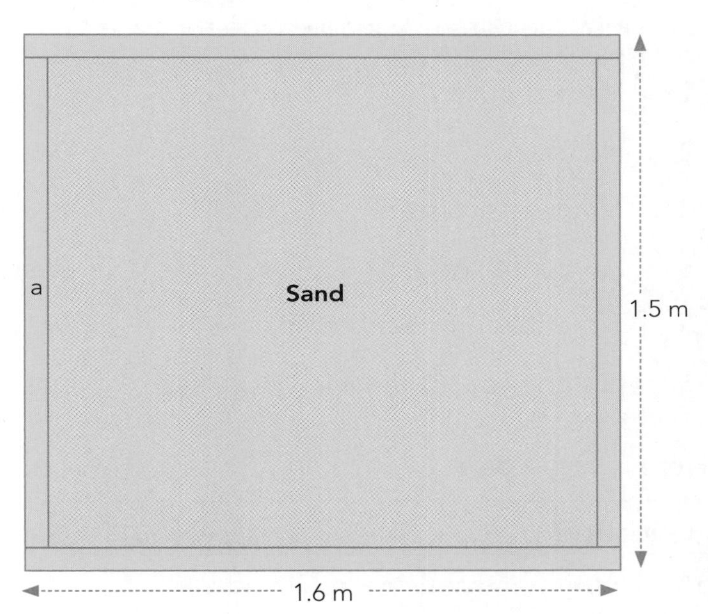

a Calculate the total area occupied by the sandpit.

Area = ________ m^2

b The four boards forming the frame of the sandpit will each be 5 cm thick. Calculate the length of board a.

Length = ________ m

c The sand will be 20 cm deep. Which is the correct calculation for Eric to use in order to calculate the volume of sand needed to fill the box?

☐ $1.5 \times 1.6 \times 20\ m^3$ ☐ $1.5 \times 1.6 \times 0.2\ m^3$

☐ $1.4 \times 1.5 \times 20\ m^3$ ☐ $1.4 \times 1.5 \times 0.2\ m^3$

ISBN: 9780170477710

Statistics and data

Types of variables

There are three type of variables:

Descriptive variables	These are usually **words**. Examples: eye colour, type of pet.
Discrete variables	These are **numbers** which are the result of **counting**. They are whole numbers. Examples: number of pets, number of T-shirts.
Continuous variables	These are **numbers** which are the result of **measuring**. They can be fractions or decimals. Examples: height, weight, distance, time.

Highlight the correct variable type for each of these.

1	Hair colour	Descriptive	Discrete	Continuous
2	Height	Descriptive	Discrete	Continuous
3	Number of sisters	Descriptive	Discrete	Continuous
4	Favourite song	Descriptive	Discrete	Continuous
5	Time taken to eat breakfast	Descriptive	Discrete	Continuous
6	Number of students in your class	Descriptive	Discrete	Continuous

Here is some data:

Question		**James**	**Lui**	**Michelle**	**Josiah**
A	How many cousins do you have?	9	2	6	3
B	What is your favourite ice cream flavour?	Chocolate	Hokey Pokey	Raspberry	Vanilla
C	How long does it take you to get to school?	20 min	12 min	3 min	27 min
D	How many cars does your family have?	1	0	3	2
E	What is your pet's name?	Rigby	Max	Angel	Rudi
F	What mass is your pencil case?	135 g	208 g	0 g	78 g

7 Which questions have answers that are descriptive variables? ____________

8 Which questions have answers that are discrete variables? ____________

9 Which questions have answers that are continuous variables? ____________

ISBN: 9780170477710

Data display

Pie graphs

- Pie graphs are appropriate for **descriptive** data. They can be used for discrete data, but usually this is not the best way of showing it.
- Pie graphs are best used when there are relatively few divisions of the data.
- The area of each sector is proportional to the frequency of each variable.

Understanding pie graphs

Examples:

1 Students were asked which language option they would like to take. Here are the results:

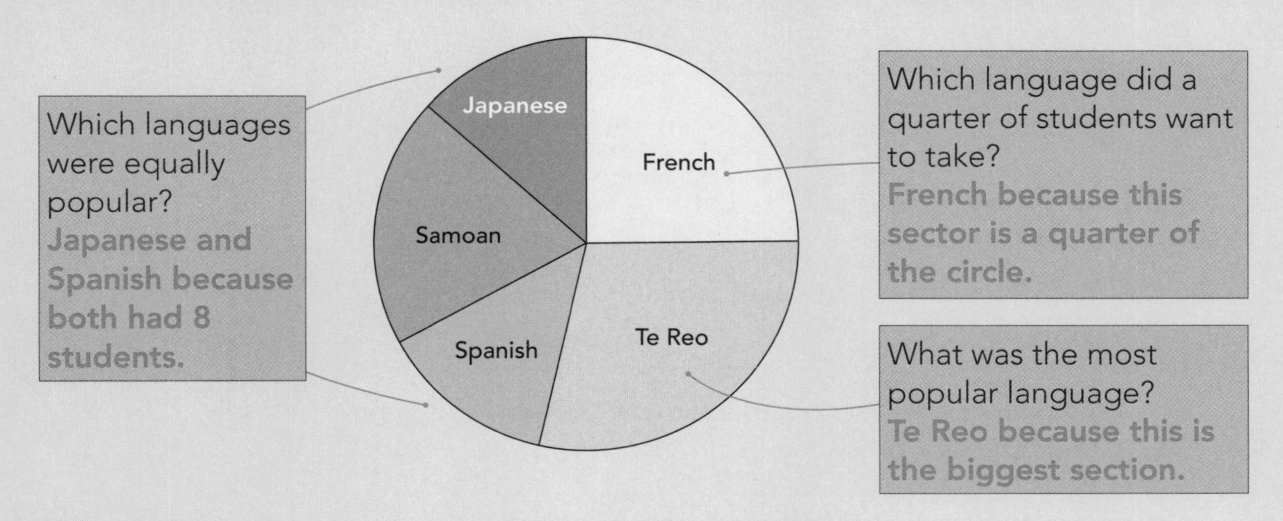

2 Forty-eight students were surveyed about their preferred pet.
In this case, numbers aren't included in the sectors so a calculation is required.

Step 1: Count how many sections there are. There are 24 sections.

Step 2: Divide the number surveyed by the number of sections: $\frac{48}{24} = 2$

Number of students

Number of sections

$\therefore$ Each section represents **2** students.

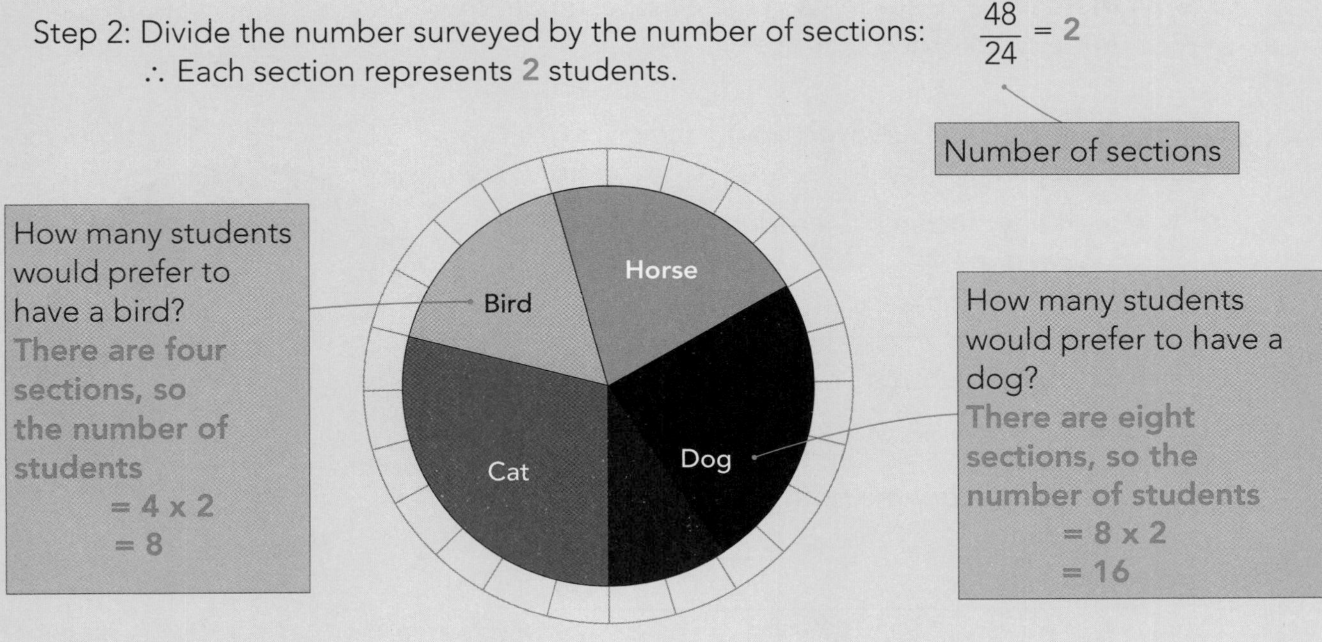

N

Answer the following questions.

1 Angus was tidying his T-shirt drawer and counted how many he had of each colour.

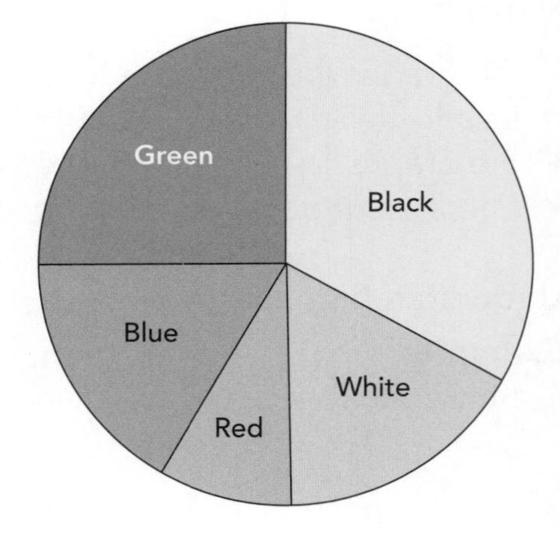

a Which of these colours was the least common?

____

b A quarter of his T-shirts are the same colour. What colour?

____

c He has equal numbers of two colours. What colours are they?

____

d Estimate the fraction of his T-shirts that are not black? ______________________

2 Members of a class were asked what they would like for lunch.

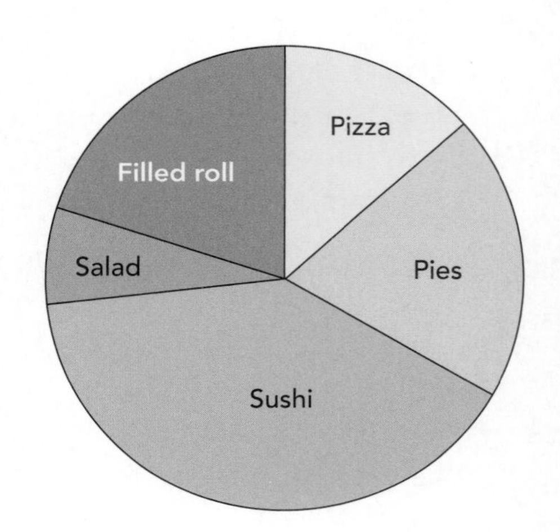

a What was the most popular choice?

____

b Which food was not chosen by more than 90% of the students?

____

c Two choices were equally popluar. What were they?

____

d If there were 30 students in the class, estimate the number who ordered sushi. ______________________

3 Eighty students were surveyed about their favourite subject.

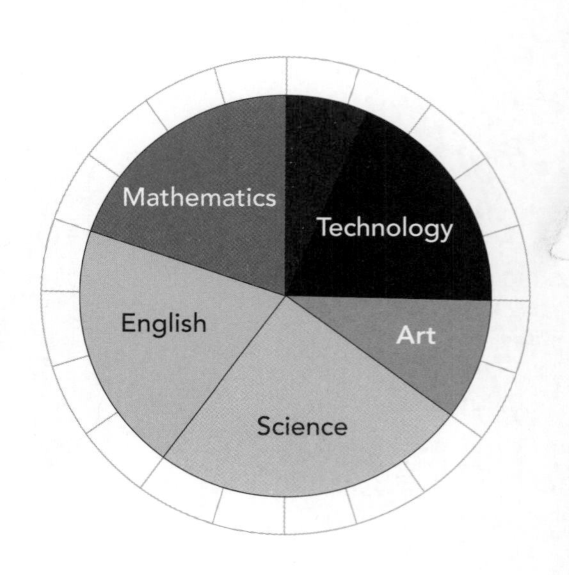

a How many students are represented by each section?

____

b How many students prefer Art?

____

c Which subjects were both preferred by 20 students?

____

d What percentage of students preferred English or Mathematics? ______________________

ISBN: 9780170477710

Strip graphs

- Strip graphs can be used with **descriptive** or **discrete** numeric variables.
- They take the form of a rectangular strip which is divided into parts that are different colours or shades.
- The **length** of each part is **proportional to the frequency** of each variable.
- Sometimes they are divided into equal sections, with each section representing one person/value, unless you are told otherwise.

Examples:

1 A group of people were asked what flavour pizza they would like.

Number of people asked = **4 + 3 + 5**
= **12**

2 Ria spent her money at the mall. Each section represents $15.

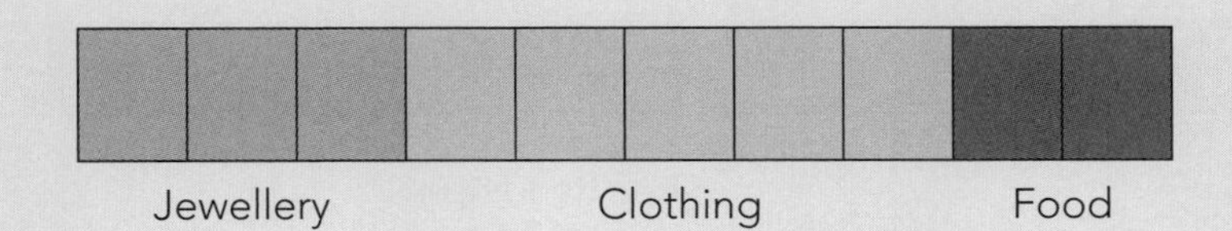

a How much money did she spend on clothing? **Amount = 5 x $15 = $75**

b How much money did she spend in total? **Amount = 10 x $15 = $150**

3 Huia went on a harbour cruise and kept a record of the types of animals she saw.

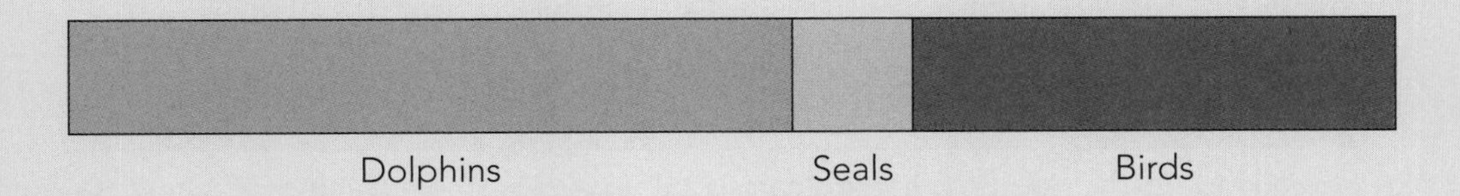

a Which animal did she see the most? **Dolphins (it's the biggest section)**

b Which animal did she see four times as much as seals? Use a ruler to check your answer. **Birds**

N

Answer the following questions.

4 Lucas asked 15 of his classmates which of three spreads (jam, honey, peanut butter) they preferred on toast. He put the information in a strip graph.

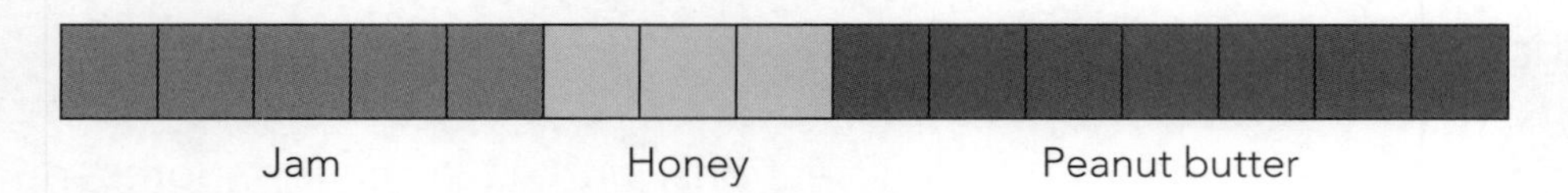

a How many preferred peanut butter or honey? ____________

b Which spread was preferred by a third of students? ____________

5 Students were asked to donate a gold coin or coins to charity. Below are the amounts they contributed.

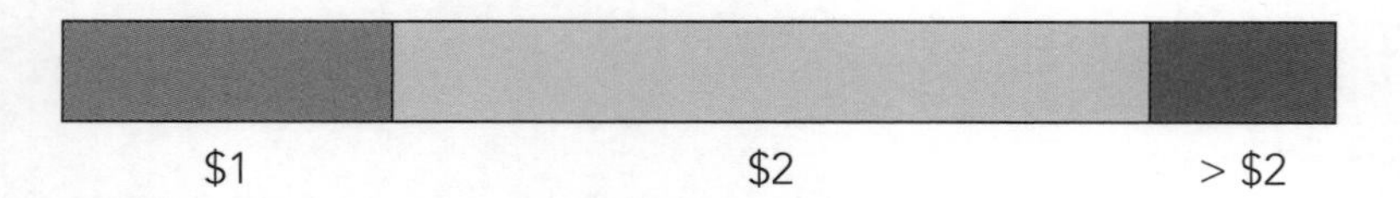

a What was the most common amount? ____________

b What does '> $2' mean? ______________________________

6 Students were asked which of five activities they spent the most time doing between 4 pm and 10 pm yesterday. Each section represents the choices of two students.

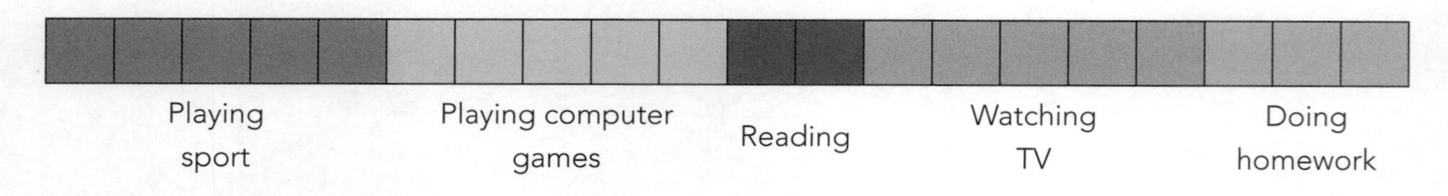

a How many students spent the most time playing sport? ____________

b How many spent the most time playing computer games or watching TV? ____________

c What percentage spent the most time doing homework? ____________

7 Gia kept track of her holiday job earnings. Each section represent $15.

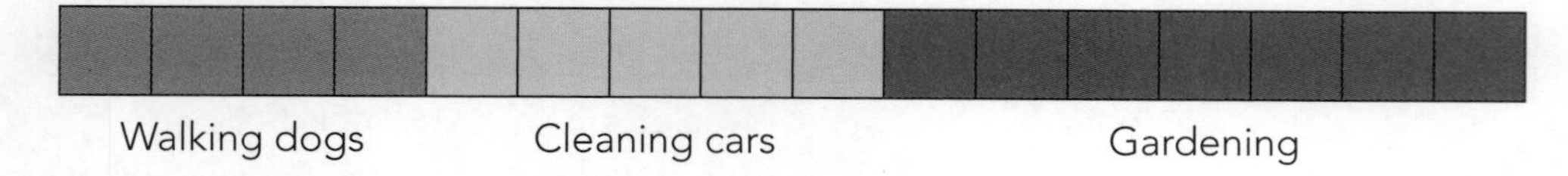

a Which job earnt her the most? ____________

b Which activity earnt her $75? ____________

c How much did Gia earn in total? ____________

 ISBN: 9780170477710

Bar graphs

- Bar graphs are used to display **discrete** or **descriptive** data.
- They are sometimes known as column graphs.
- They can be plotted **vertically** or **horizontally**.
- The bars always have **gaps** between them.

Examples:

1 Data was collected on students' favourite fruits.

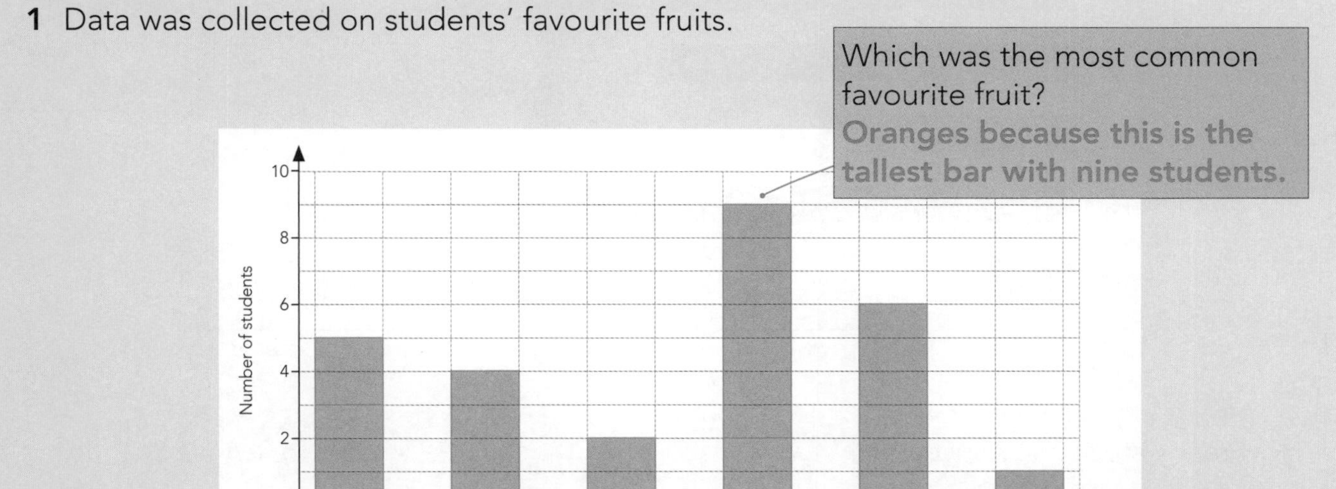

2 Several groups can be plotted on the same graph. Favourite vegetables were collected from two year groups.

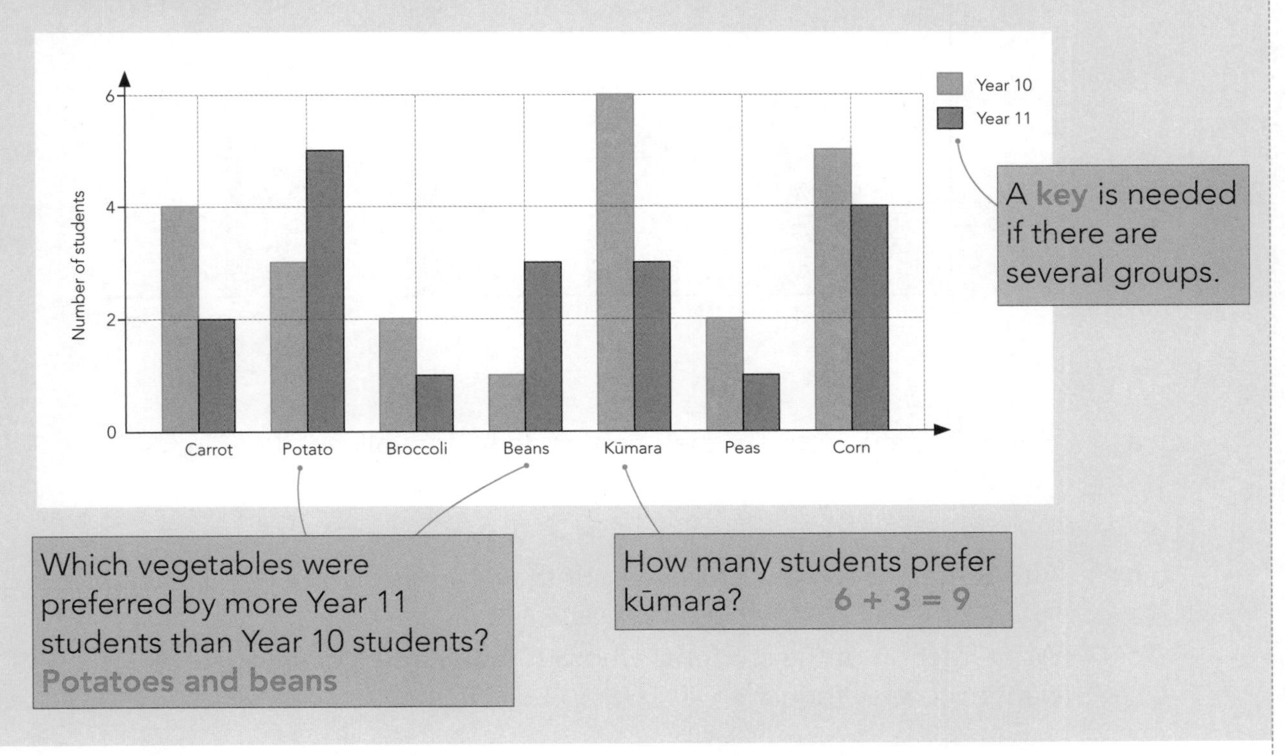

N

3 Bar graphs can also be displayed horizontally.
Students were asked to respond to this statement:
'I enjoy having a longer lunchtime.'

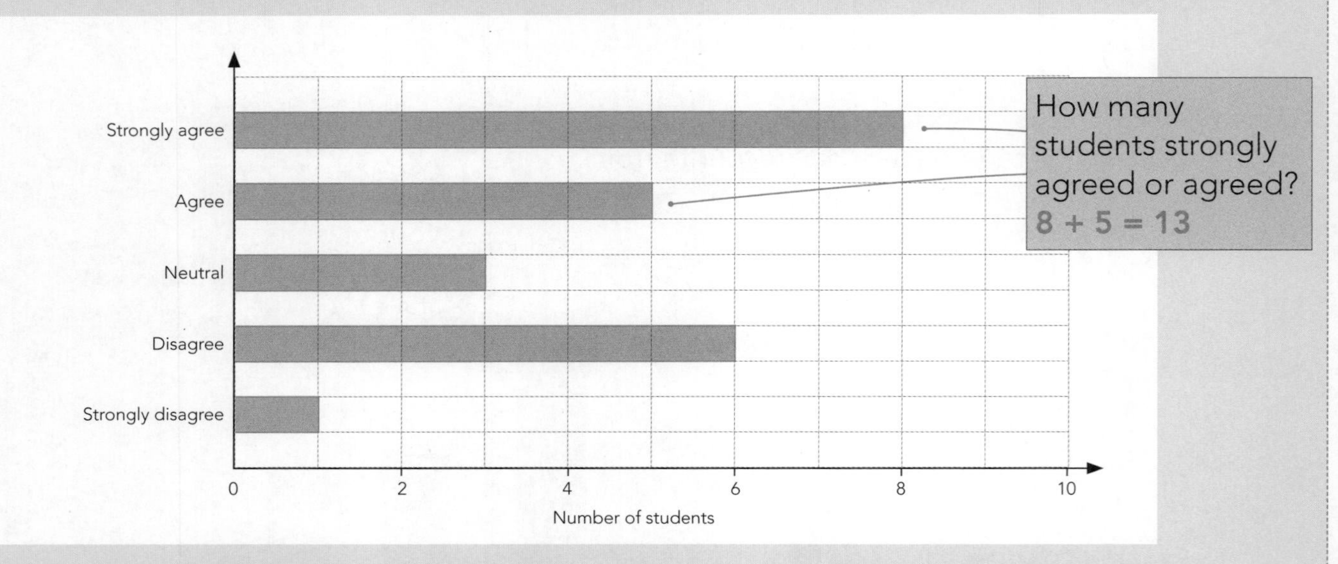

How many students responded? **8 + 5 + 3 + 6 + 1= 23**

Answer the following questions.

8 Students put their names down for a summer sport. The results are shown in the graph.

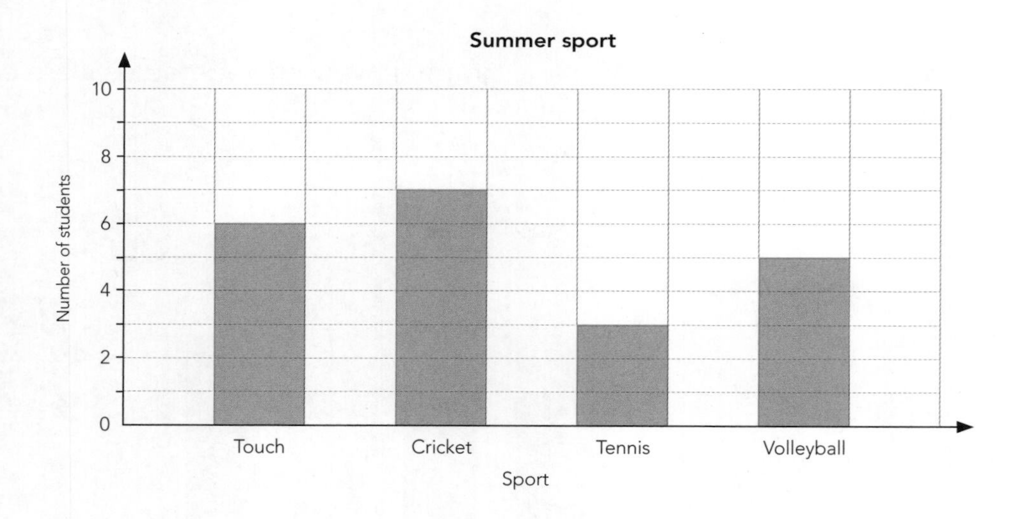

a How many students put their names down for volleyball or tennis? ____________

b Altogether, how many students put their names down? ____________

c What fraction of the students who put their names down wanted to play tennis? ____________

ISBN: 9780170477710

9 This graph shows the numbers of votes for the five most popular New Zealand birds of the year in 2021. These five got 19 502 votes altogether. (Pekapeka were included to raise awareness of this flying mammal.)

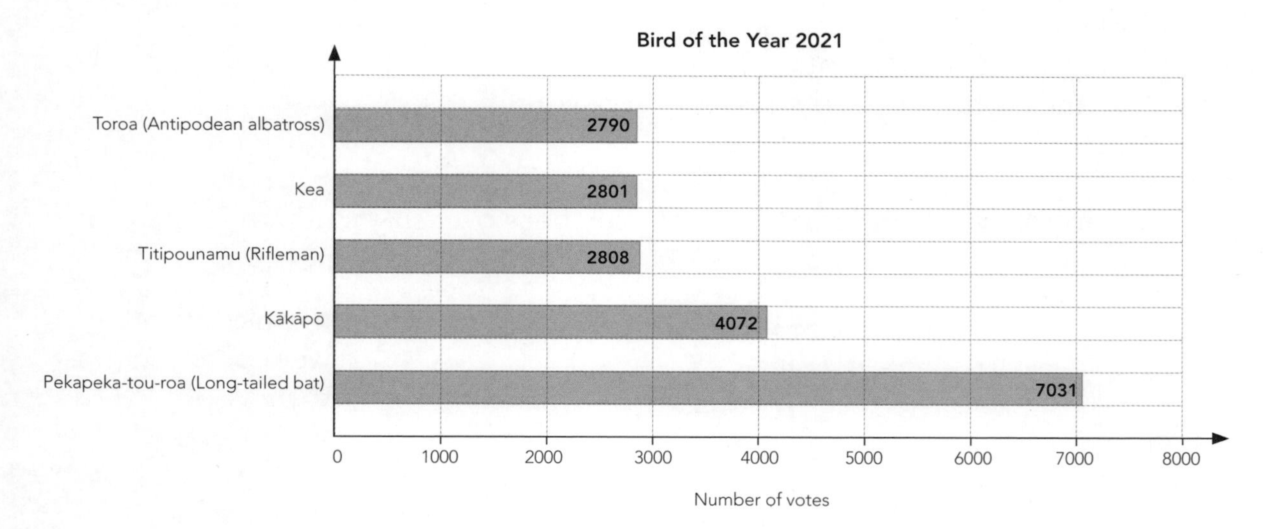

a What percentage of these votes went to the toroa? Give your answer to 1 dp. __________

b Altogether, how many votes did the kea and the titipounamu receive? __________

c How many more votes would the kākāpō have needed to win? __________

10 The graph shows the popularity of some baby names in New Zealand.

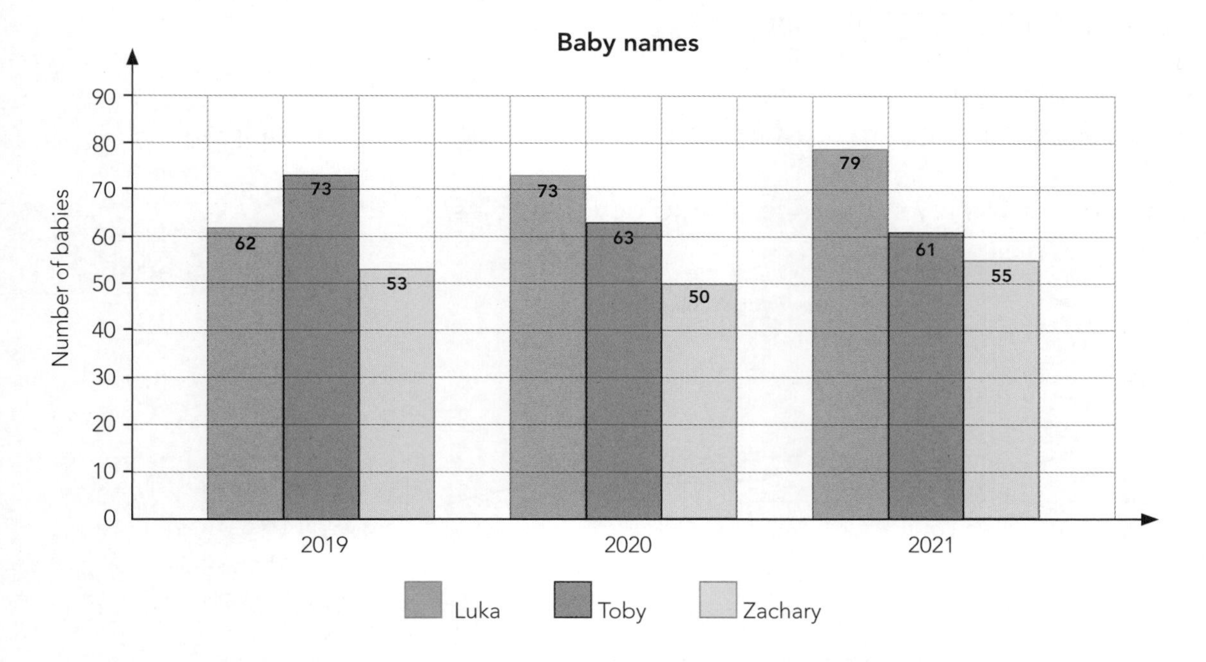

a Which of the three names was most popular in 2019? __________

b During the three years, how many babies were named Luka? __________

c Which name decreased in popularity over the three years? __________

ISBN: 9780170477710

Line graphs

- Line graphs are often used to show how **discrete** or **continuous** data changes at regular intervals of time.
- Lines connect the plotted points.
- Sometimes these are called **time series** graphs.

Examples:

1 The line graph shows the price of 1 kg of frozen peas over six months in 2022.

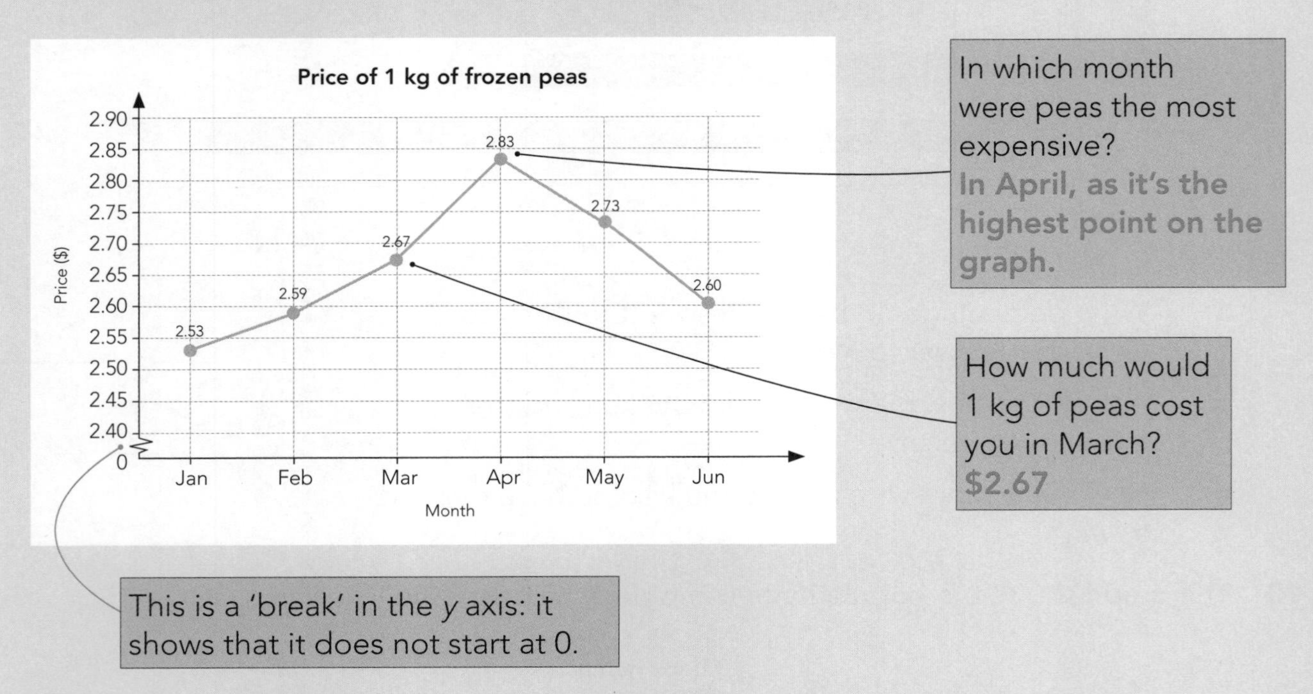

2 There can be several lines on one graph.
Sometimes line graphs are used for comparison rather than for reading values.

The graph below shows the price of 1 kg of vegetables.

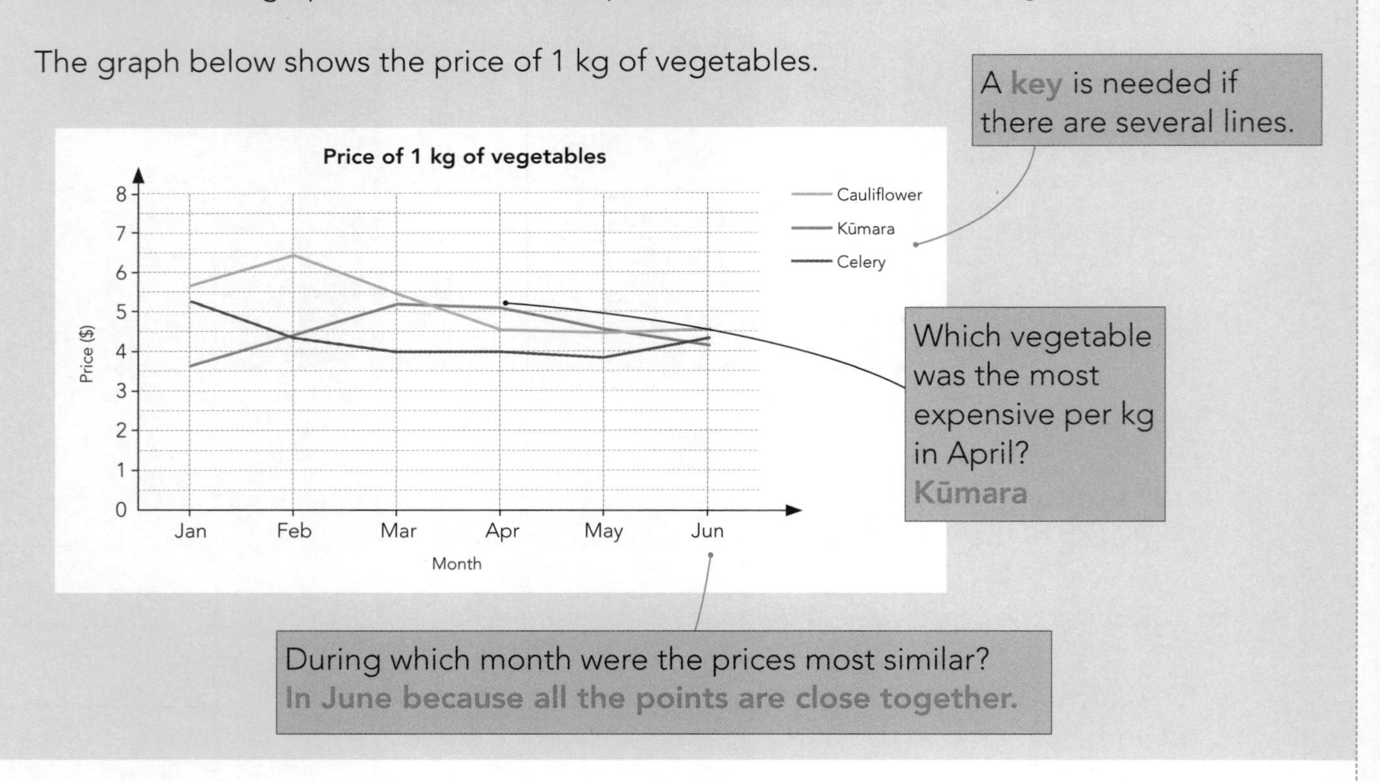

ISBN: 9780170477710

Answer the following questions.

11 This graph shows the mean monthly temperatures in Whangārei.

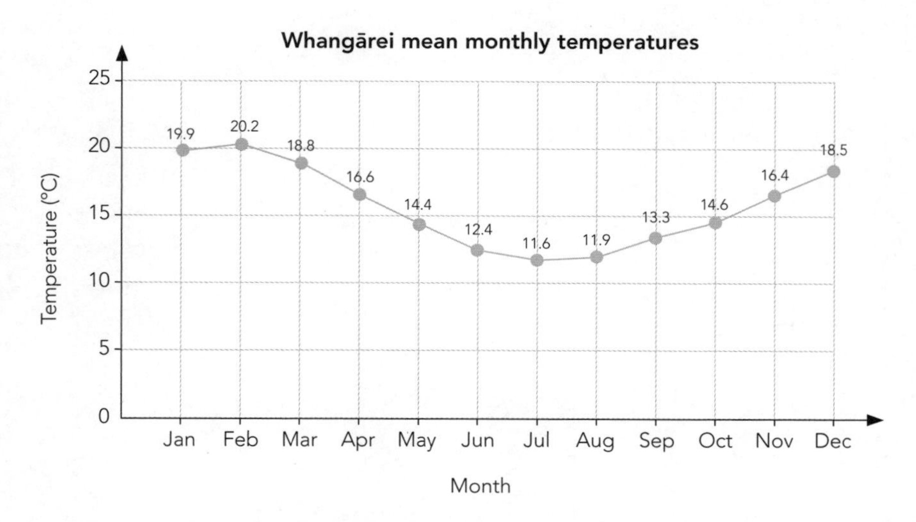

a What is the mean temperature in October? ________________

b Which two months had the coldest mean temperatures? ________________

c How many months have an average temperature over 15°C? ________________

12 The graph shows the number of male and female athletes who represented New Zealand at the summer Olympics between 1984 and 2020.

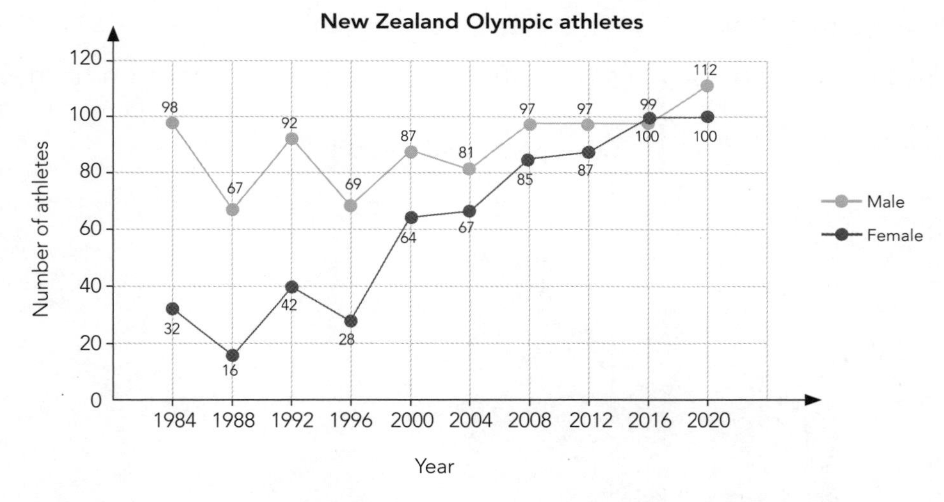

a How many of our male athletes went to the 1996 Olympics? ________________

b In which year were there more females than males athletes? ________________

c How many athletes went to the 2012 Olympics? ________________

d In which year did we send the smallest group of athletes? ________________

ISBN: 9780170477710

N

Histograms

- Histograms are used to display **continuous** (**measured**) data.
- The data is displayed in **intervals**.

Example:

The arm spans of students are recorded below.

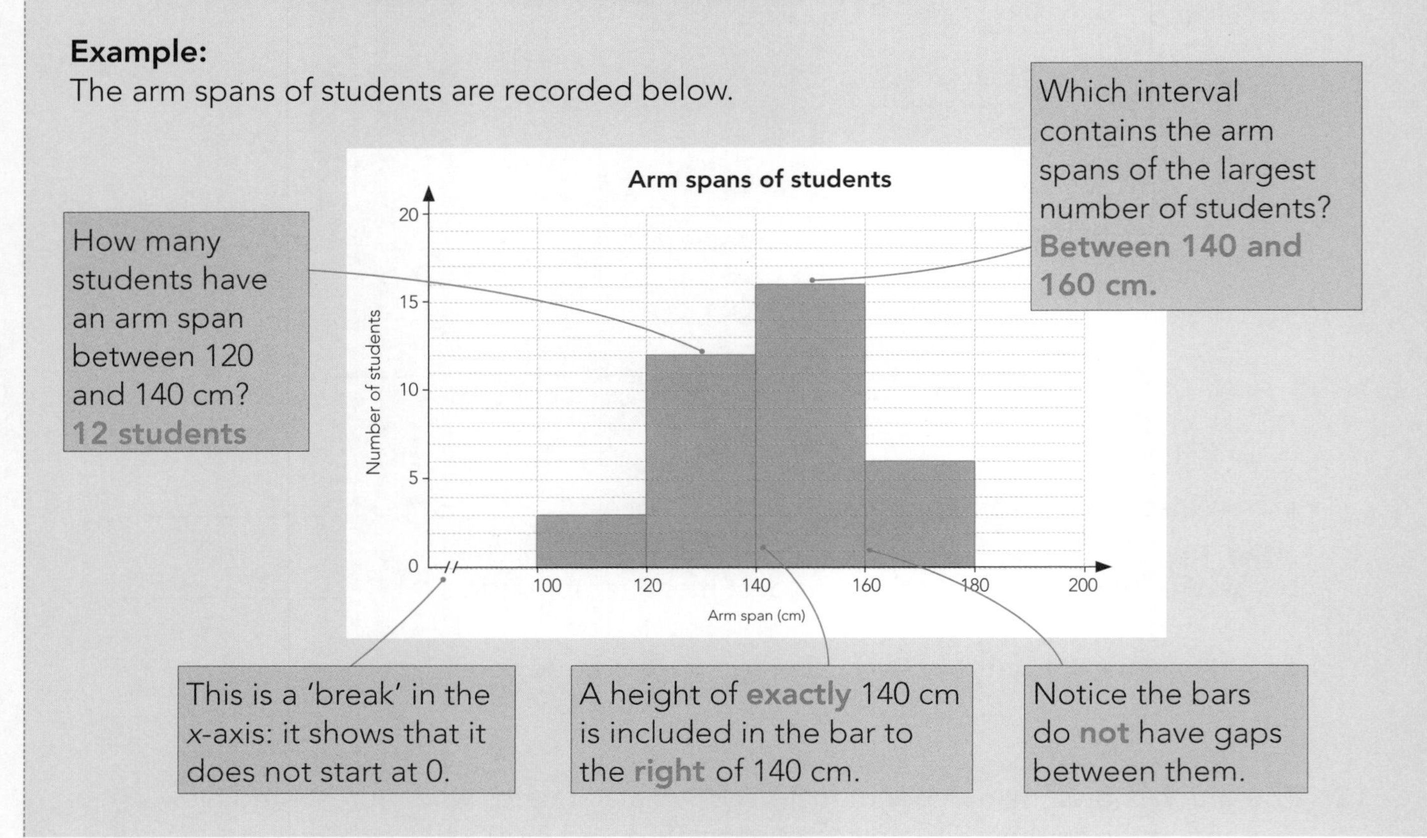

Answer the following questions.

13 Javier asked the members of his year group how long they spent on homework last night.

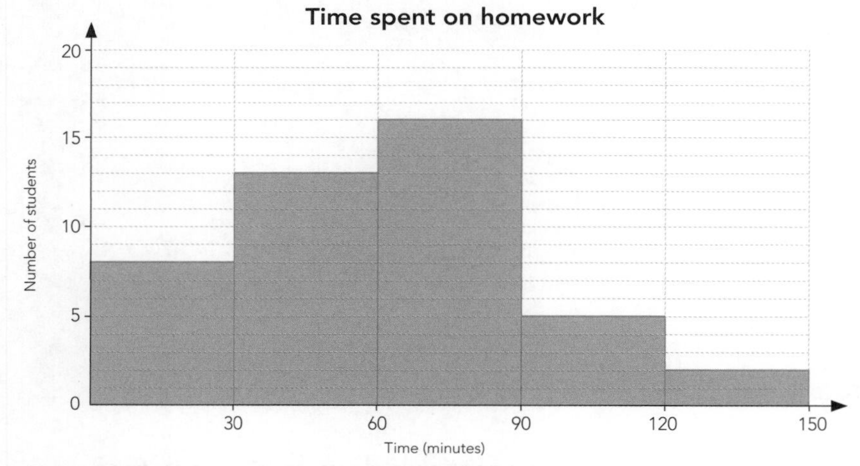

a How many students spent between 30 and 60 minutes? ________

b How many students spent 90 minutes or longer on homework? ________

c What fraction of the year group spent less than 30 minutes on homework? ________

ISBN: 9780170477710

14 This graph shows the percentage of battery remaining on the students' cellphone.

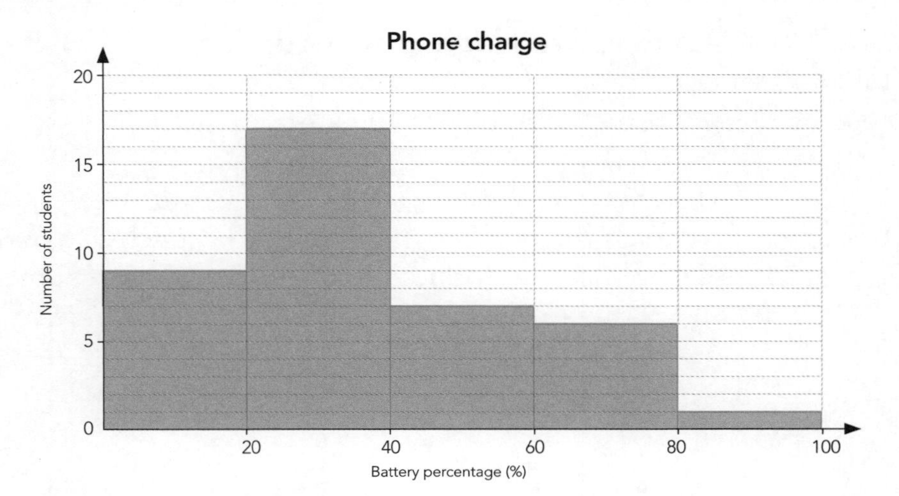

a How many students have between 40 and 60% charge left on their phone? ____________

b How many students have less than 40% charge left on their phone? ____________

c What is the least common amount of battery charge? ____________

d How many students were asked about their phone battery life? ____________

15 This graph shows the screen time for a group of students from the previous day.

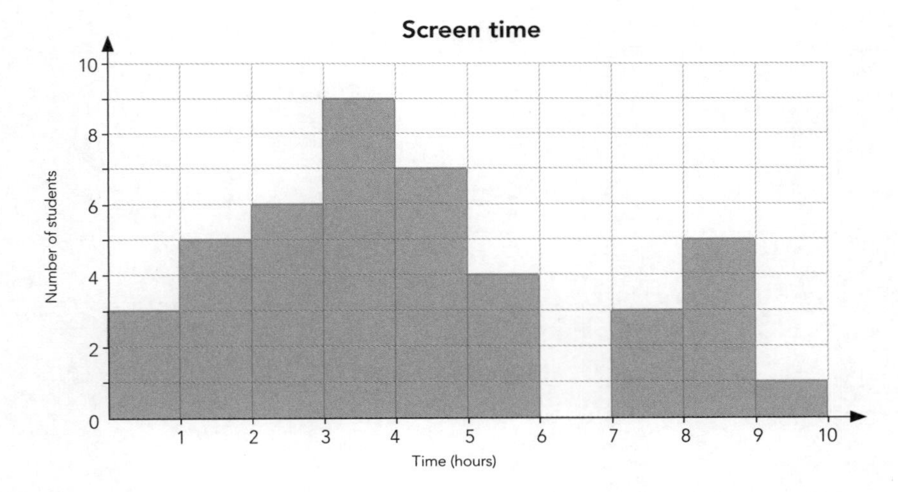

a How many students had a screen time between 4 and 5 hours? ____________

b What was the most common amount of screen time? ____________

c Why is there a bar missing? ____________

d How many students spend at least seven hours watching a screen? ____________

N

Dot plots

- Dot plots are used for **discrete** and **rounded continuous** data.
- They are useful for comparing groups.
- Each dot represents one person/object, unless you are told otherwise.

Examples:

1 Hemi keeps chickens. He records the number of eggs they lay each day during December. He drew a dot plot to show the number of days on which he got each number of eggs.

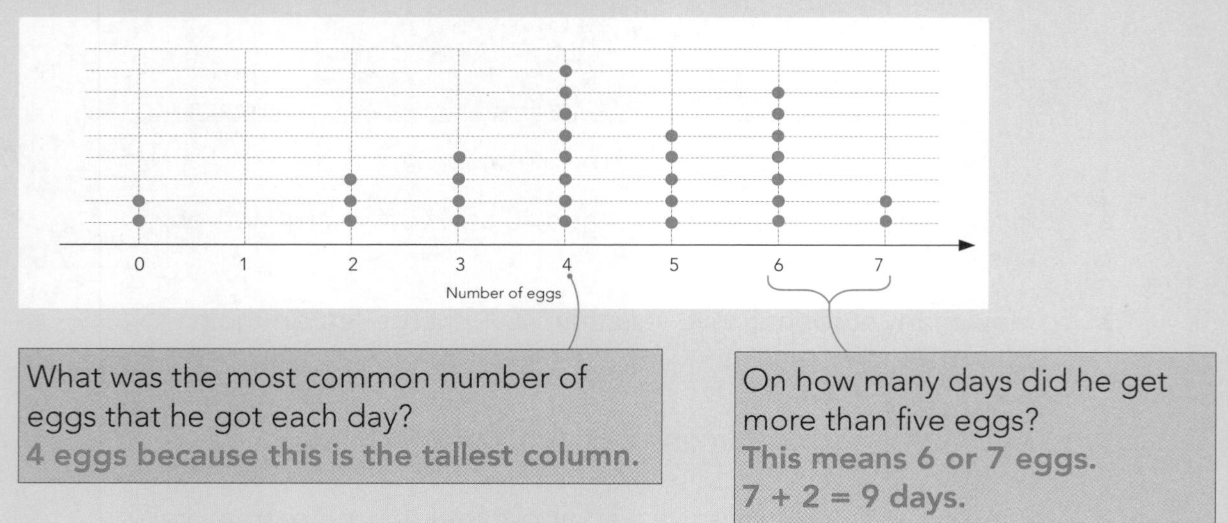

Dot plots are useful for comparing data sets.

2 Two classes plotted the results of their recent quiz scores.

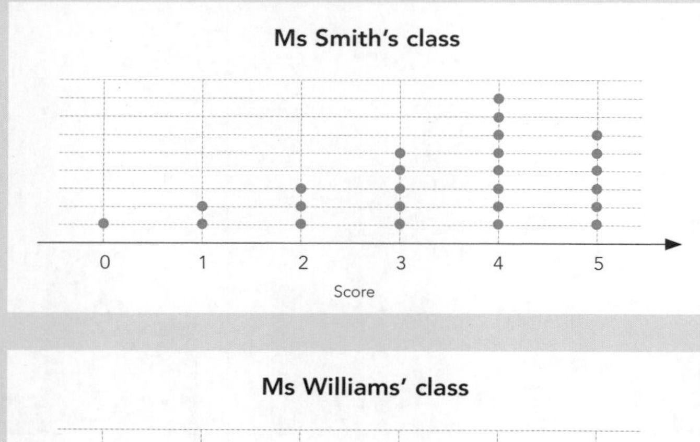

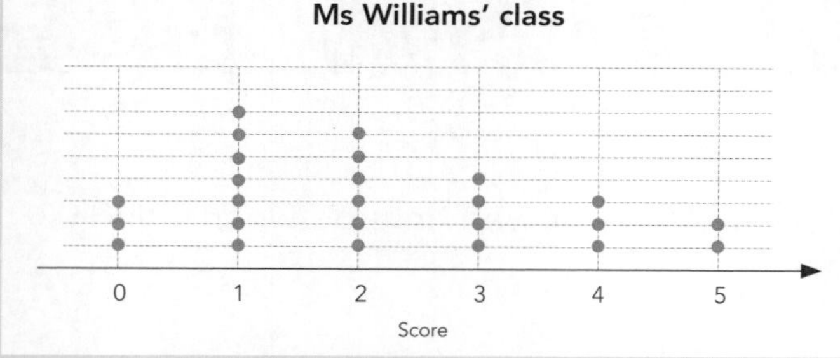

Which class scored better in the quiz? **Ms Smith's class**

Explain your answer. **There are more dots towards the right for Ms Smith's class. This means that more students in her class scored better in the quiz.**

 ISBN: 9780170477710

Answer the following questions.

16 Moana asked all the students in a class which of Mars, Moro, Picnic, Crunchie, Snickers and KitKat they liked best.

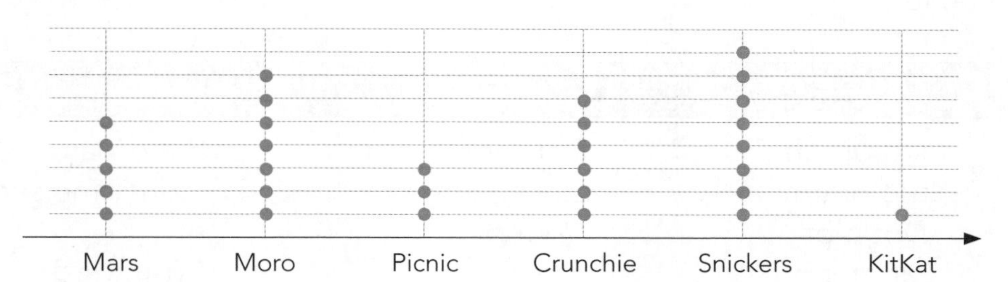

a How many students were in the class? ____________

b Which chocolate bar was preferred by only five students? ____________

c Moana is organising the class party and can buy only three types of chocolate bar. Which three should she buy? ____________

d What fraction of the class preferred Mars or Moro bars? ____________

17 The Year 11 and Year 12 Agriculture classes held a challenge. Each student in the two classes was given one seed potato to plant in the school garden. All the potatoes were harvested on the same day and, for each plant, the number of potatoes with a mass of at least 40 g was counted. Here are the results:

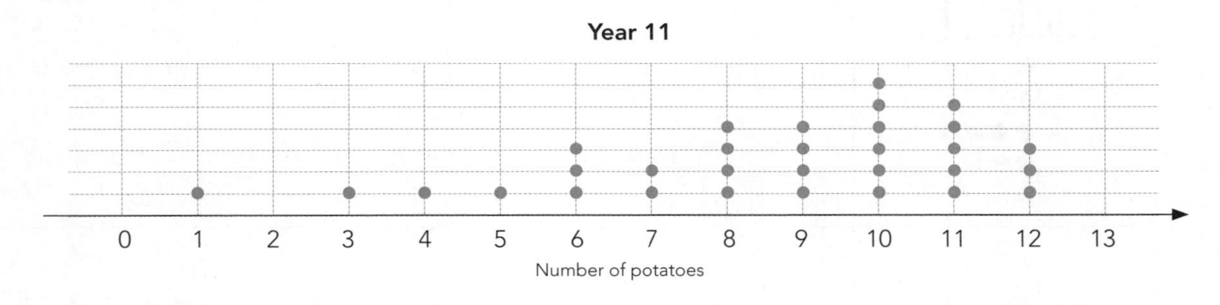

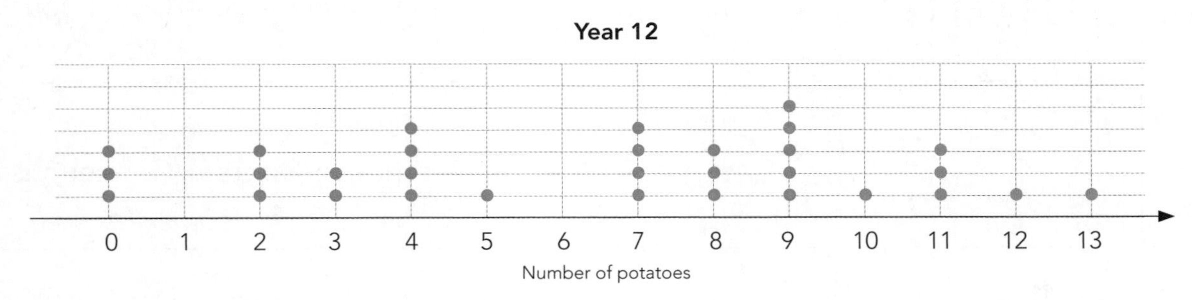

a Which class did better? ____________

b Explain your answer. ____

____

c How many students in total grew more than 11 potatoes with mass over 40 g? ____________

ISBN: 9780170477710

N

Summary

- There are different ways of displaying data. Which is most appropriate depends on the **type** of variable.

Descriptive	Discrete data	Continuous data
Pictograph Strip graph Dot plot Bar graph Pie graph	Pictograph Strip graph Dot plot Bar graph Line graph	Histogram Line graph/Time series **If rounded:** Dot plot

Note: A pie graph can be used to plot discrete data, but a bar graph is a much better choice.

18 Complete the table.

Graph	Name	Suitable for
	Pie graph	Descriptive/Discrete/Continuous data
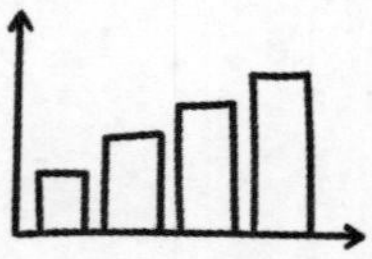		Descriptive/Discrete/Continuous data
Jan Feb Mar Apr		Descriptive/Discrete/Continuous data
0 1 2 3 4 5 6 7 8 9		Descriptive/Discrete/Continuous data
0 10 20 30 40 50 60 70 80 90 Oct 2019 Nov Dec Jan 2020 Feb Mar Apr May Jun		Descriptive/Discrete/Continuous data
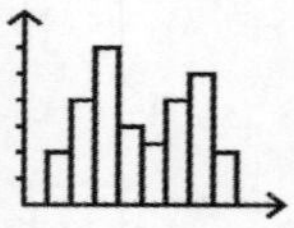		Descriptive/Discrete/Continuous data
Soccer Basketball Netball Hockey Rugby league Rugby union		Descriptive/Discrete/Continuous data

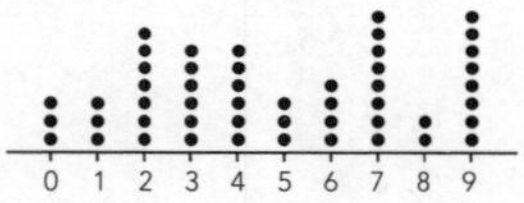

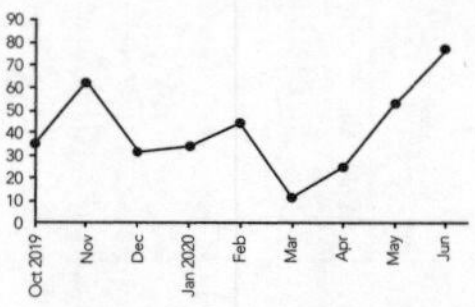

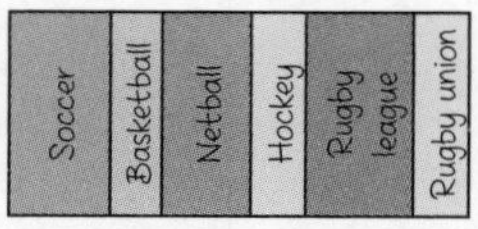

ISBN: 9780170477710

Data analysis

- There are two measures that we need to know in order to be able to discuss and compare distributions:

  1 Where is the **centre** of the data?
  2 How widely is the data **spread**?

Measures of centre (averages)

- There are **three** measures for the centre of data.

Name	Calculation
Mean	$\frac{\text{the sum of all the data values}}{\text{the number of data values}}$
Median	middle value
Mode	value that occurs most frequently

Mean

- The mean is sometimes falsely called **the** average. It is **an** average.
- The numbers do not need to be in order for this calculation.
- Means are often long decimals, so sensible rounding may be needed.
- The mean is influenced by unusually large or small values.

$$\text{mean} = \frac{\text{sum of all the data values}}{\text{number of data values}}$$

Examples:

1 8 1 7 4 6 8 3 5 3

$$\text{Mean} = \frac{8 + 1 + 7 + 4 + 6 + 8 + 3 + 5 + 3}{9}$$
$$= 5$$

There are 9 numbers in the data set.

The **mean** of this data set is **5**.

2 9 2 1 4 6 0 3 5

$$\text{Mean} = \frac{9 + 2 + 1 + 4 + 6 + 0 + 3 + 5}{8}$$
$$= 3.75$$

Notice that 0 must be included in the calculation.

The **mean** of this data set is **3.75** or **3.8** (1 dp).

Calculate the means of these data sets.

1 **a** 6 4 2 8 9 1 3 7 9 3 Mean = $\frac{6 + \_\_\_\_\_\_\_\_}{10}$ = ______

b 5 0 7 3 4 2 9 1 Mean = $\frac{\_\_\_\_\_\_\_\_}{8}$ = ______

c 12 11 16 19 14 12 Mean = ____________ = ______

d 25 31 27 20 42 31 27 Mean = ____________ = ______

N

Median

- If there is an **odd number of values** in a data, the median is the **middle number**.
- If there is an **even number of values**, the median is **halfway between the two middle numbers** in the data set.
- Before you can calculate the median, you must **put the data in order**.

Examples:

1 A data set with an odd number of values

1 2 4 4 6 7 8 8 9

This is the middle number.

The median of this data set = **6**

2 A data set with an even number of values

1 3 5 7 9 10 11 13

These are the middle numbers. Add them together and divide by 2.

The median for this data set = $\frac{7 + 9}{2}$ = **8**

3 An unordered data set

~~2 7 3 6 1 2 7 8 3 9~~

Cross them off as you go, to make sure you don't miss any.

Put them **in order** before finding the median.

1 2 2 3 3 6 7 7 8 9

The median for this data set = $\frac{3 + 6}{2}$ = **4.5**

2 Find the medians of these data sets.

a 0, 1, 1, 2, 3, 3, 4, 5, 7, 7, 8, 9, 10 Median = ________

b 5, 7, 7, 11, 15, 16, 17, 19, 19, 21, 23, 26 Median = ________

c 3, 5, 5, 7, 9, 10, 10, 13, 15, 17, 21, 23, 27, 28 Median = ________

d

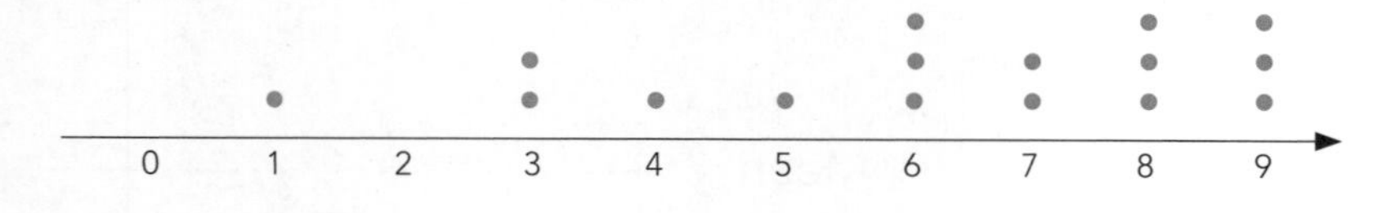

Median = ________

e

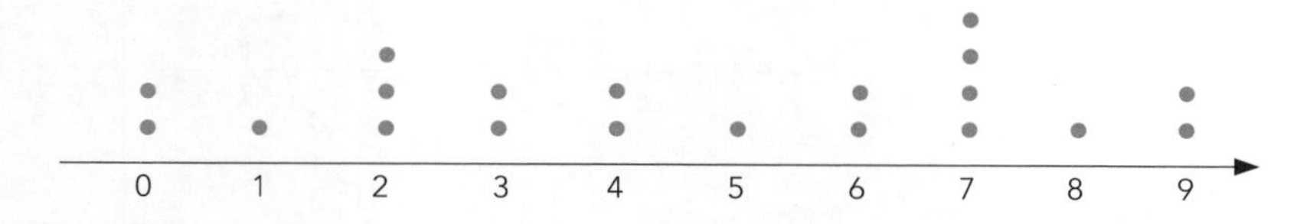

Median = ________

ISBN: 9780170477710

Mode

- The mode is the **most common value**.
- Sometimes there are **several modes**.
- If there are **three or more** numbers that occur equally often, we say there is **no mode**.

Examples:

1

The most common number is 2: there are four of them.

The mode of this data set is 2.

2

Both **1** and 6 occur three times.

The modes are 1 and 6.

3 5 8 4 2 3 8 5 1 2

There are three numbers that occur equally often: **2**, 5 and **8**.

If there are **three or more** modes, we say the data is **polymodal**.

3 Find the mode(s) of these data sets.

a 6 2 1 4 6 2 5 9 2 1 Mode(s) = ______________

b 10 8 7 4 9 7 4 1 5 6 Mode(s) = ______________

c 7 3 0 1 4 3 6 1 7 1 0 3 7 4 Mode(s) = ______________

d

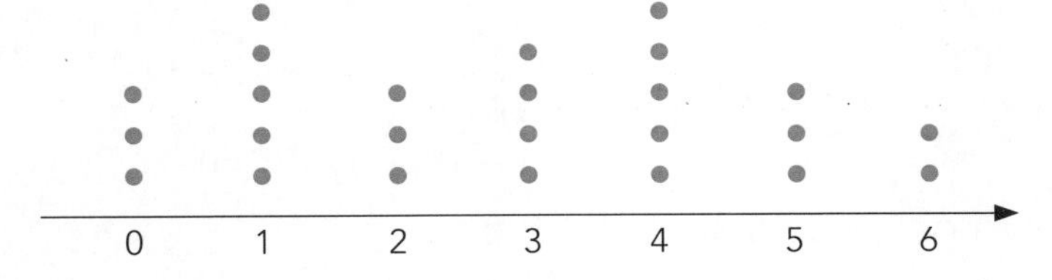

Mode(s) = ______________

e

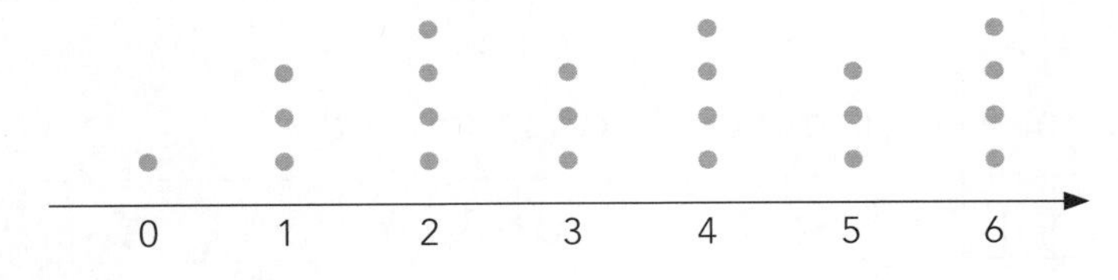

Mode(s) = ______________

f

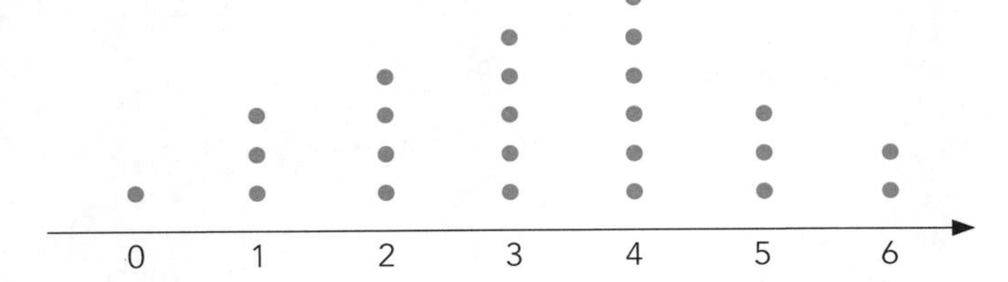

Mode(s) = ______________

ISBN: 9780170477710

N

Measure of spread — range

- The range is the **maximum value minus the minimum value** in the data set.
- Note: the range is a **single number**.
- Like the mean, the range is affected by unusually large or small values.
- The data does not need to be in order to calculate the range.
- The range is a measure of the **variability** of the data.
- Calculating the range for several data sets can help us decide **which is more spread out**.

Range = maximum – minimum

Examples:

1

12 3 10 6 4 6 9 15 16 4 8 11

The smallest value in the data set (minimum).

The largest value in the data set (maximum).

Range = 16 – 3
= 13

The range or the spread of this data set is 13.

2 Two classes took the same test this week. There were 10 questions. The table shows the results.

11A	4	7	8	3	9	10	2	5	8	8	4	7	2	3	1	4	9	3	5	1
11B	8	2	7	6	6	3	7	4	4	2	5	8	9	3	3	7	4	2	9	3

11A: Range = 10 – 1
= 9

11B: Range = 9 – 2
= 7

∴ 11A has a larger range (9) than 11B (7), so the test results for 11A are more variable.

Answer the following questions.

4 Calculate the range for the following data sets.

a 7, 2, 6, 5, 4, 4, 9, 2, 3, 8, 13 Range = ________________

b 10, 17, 6, 25, 12, 16, 4, 7, 19 Range = ________________

5 **a** The largest number in a data set is 14, the range is 9. Smallest number = __________

b The smallest number in a data set is 32, the range is 61. Largest number = __________

6 Two other classes took the same test. The table shows the results.

11C	9	5	4	6	3	6	4	5	5	7	8	10	8	5	3	8	3	10	6	9
11D	5	9	5	4	4	3	2	6	8	9	4	3	1	5	3	4	7	10	6	3

Which class had more variable results?

11C: Range = _____ – _____
= ______

11D: Range = _____ – _____
= ______

∴ 11___ has a larger range (___) than 11___ (___), so the test results for 11___ are more variable.

ISBN: 9780170477710

Unusual features

- There are two types of unusual features to watch for: **unusual points** and **clusters**.
- An **unusual point** is one that is **away** from the rest of the data.
- A **cluster** is a **group** of data that is away from the rest of the data.

How many make a cluster?
Don't get hung up on labels and definitions; it's best to write what you see. If there are two points away from the rest, then say that.

Examples:

1 An unusual point.
Masses of lambs at 8 weeks (kg): 18.5, 19.1, 19.3, 18.2, 16.1, 19.8, 18.9, 18.6

This lamb's mass (16.1 kg) is 2.1 kg less than the next heaviest lamb (18.2 kg). So 16.1 kg is an unusual point.

2 It is often easiest to see unusual points and clusters when the data is graphed.

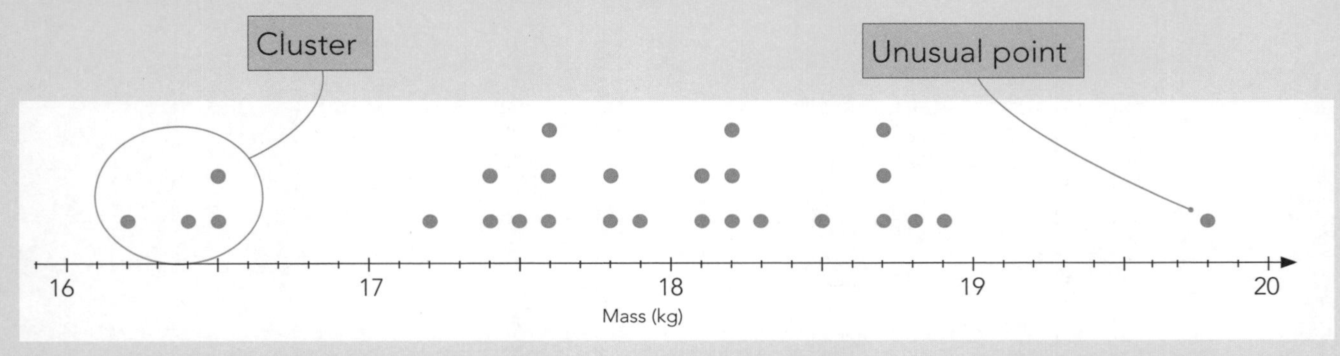

Note: **1** Check that unusual points are not mistakes in measurement, counting or recording.
2 Include them when you graph and analyse your data. Think about whether the unusual feature is possible and state your thoughts when you discuss the data.

Circle or highlight any unusual features in these data sets and state whether they are clusters or unusual points.

1

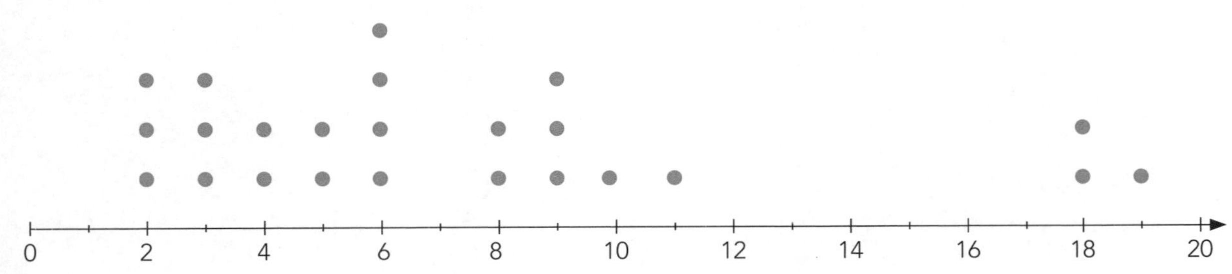

ISBN: 9780170477710

N

2

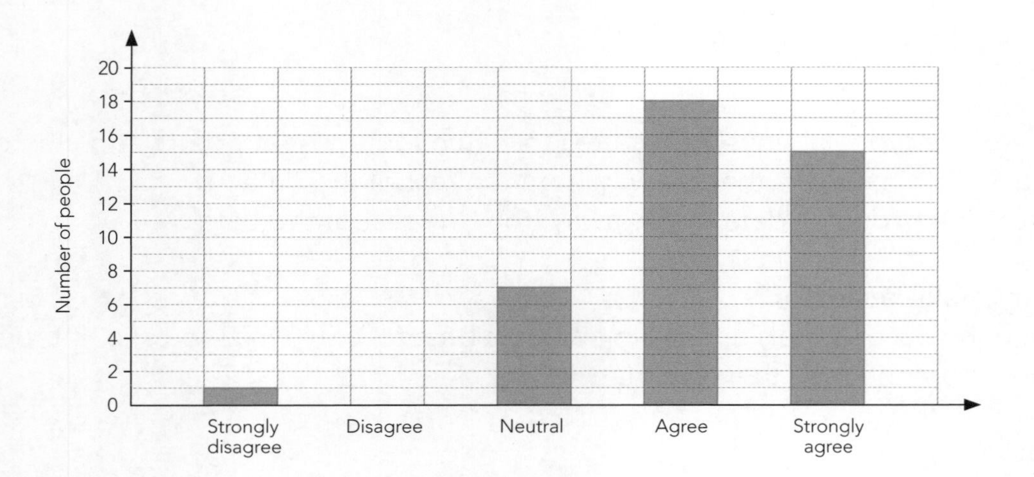

3

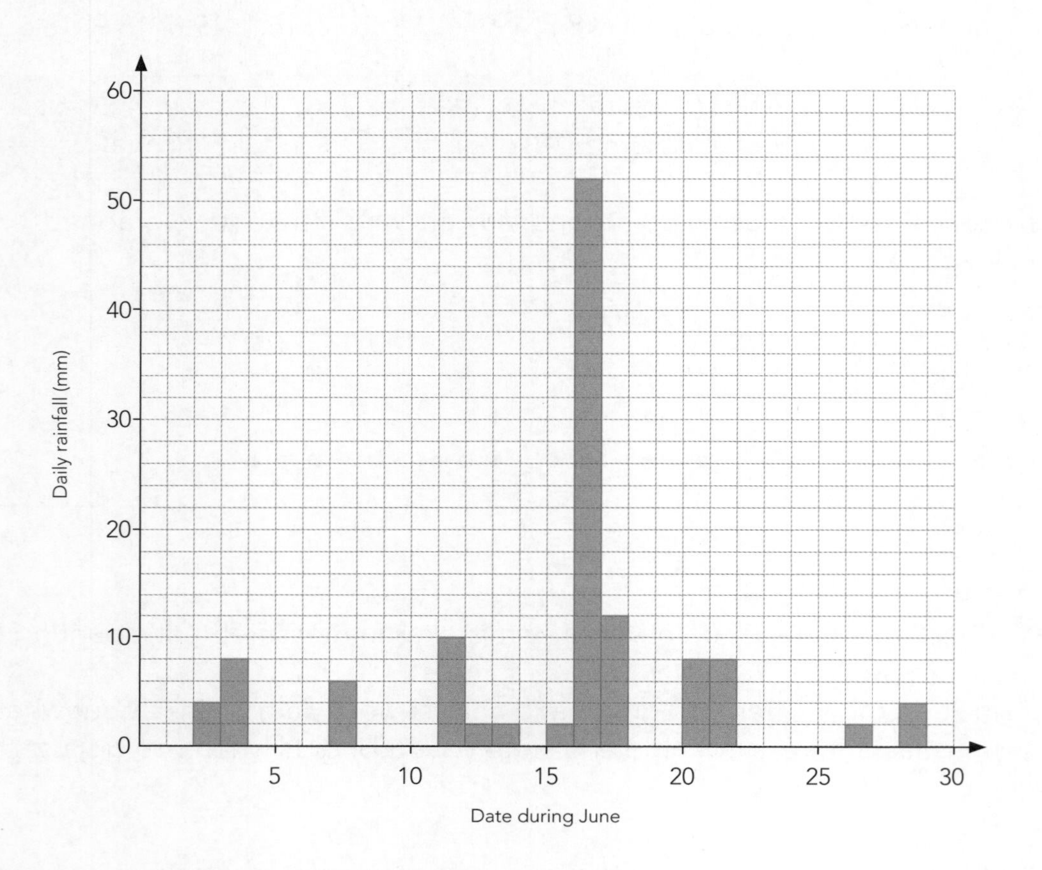

4 Test marks out of 20:

15, 18, 12, 16, 2, 19, 20, 15, 14, 3, 13, 16, 2, 14, 18

____

5 Number of siblings:

0, 3, 2, 1, 0, 1, 1, 2, 9, 2, 1, 0, 4, 2, 3

____

ISBN: 9780170477710

Elements of chance

The language of probability

- Probability is used to describe **how likely** something is to happen.
- There are many words that are used to describe probabilities.

1 Two of these terms could be used to describe similar probabilities. Highlight the word that does **not** belong with the others.

a	no way	almost certain	impossible
b	possible	slight chance	guaranteed
c	even chance	maybe	impossible
d	impossible	probable	no chance
e	maybe	certain	definite
f	no way	probable	very likely
g	certain	a sure thing	unlikely

2 Use each of these terms once only to describe the probability of each of these events occurring.

impossible **even chance** **unlikely** **likely** **certain**

a The next person you meet will be 150 years old. ______________________

b Next Friday there will be clouds in the sky. ______________________

c The Prime Minister sits next to you on the bus. ______________________

d If today is Thursday, then tomorrow is Friday. ______________________

e You toss a coin and it lands on heads. ______________________

ISBN: 9780170477710

N

The probability scale

- A probability value tells us **how likely** it is that an event will occur.
- We use numbers between **0** and **1** to describe probability.

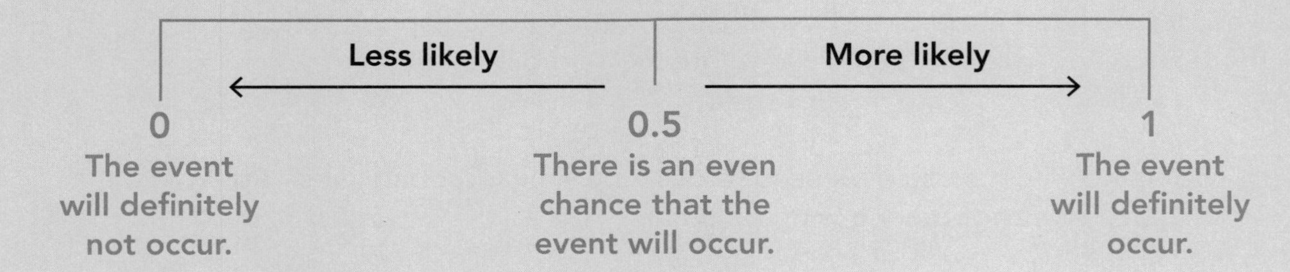

1 Discuss the meanings of these words and terms with your neighbours, and match them to their appropriate probability value. Some words may fit in several places. If you have different answers, discuss them with your teacher.

probable	very likely	slight chance	50:50	even chance	certain
very unlikely	maybe	a sure thing	likely	good chance	unlikely
guaranteed	no chance	definite	impossible	almost certain	no way

ISBN: 9780170477710

Tools commonly used in probability

- Probabilities can be written as **fractions**, **decimals** or **percentages**.
- There are a number of tools that are useful in probability.
- Unless we are told otherwise, assume coins, dice, etc. are **fair**: they produce equally likely outcomes.

Coins

- These are often used in experiments or situations when a probability of 0.5 is needed, e.g. which team gets to choose the end they play from in a game.

The probability that a coin lands on its 'head' is 0.5.

'**Heads**' is the side with the king or queen's head on it.

The probability that a coin lands on its 'tail' is 0.5.

'**Tails**' is the side that does not have the king or queen's head on it.

Dice

- We speak of **several dice**, or **one (single) die**.
- A standard die has six sides.
  There is one number on each side:
  1, 2, 3, 4, 5, 6.

  The probability of rolling each of these numbers is $\frac{1}{6}$.

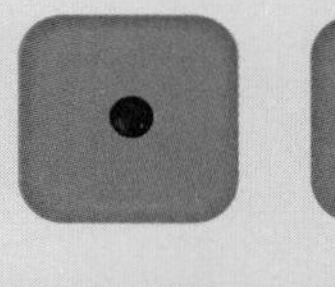
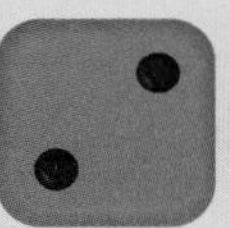

Spinners

- These are sometimes found in board games. The probability of landing on any one of the sectors depends on each spinner.

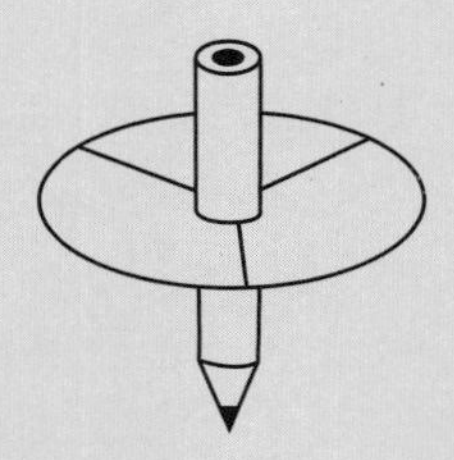

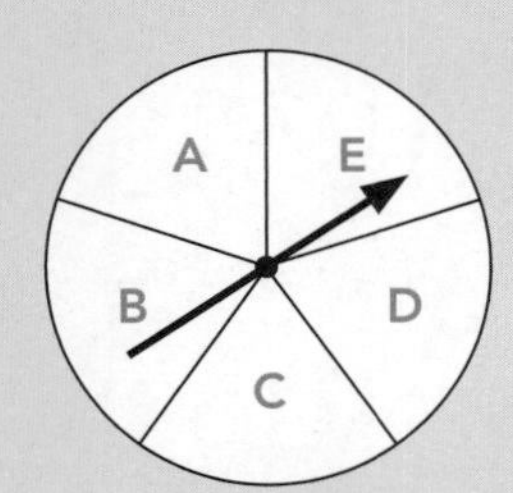

There are 5 equal sectors in this spinner, so the probability of landing on any of these sectors is $\frac{1}{5}$ or 0.2.

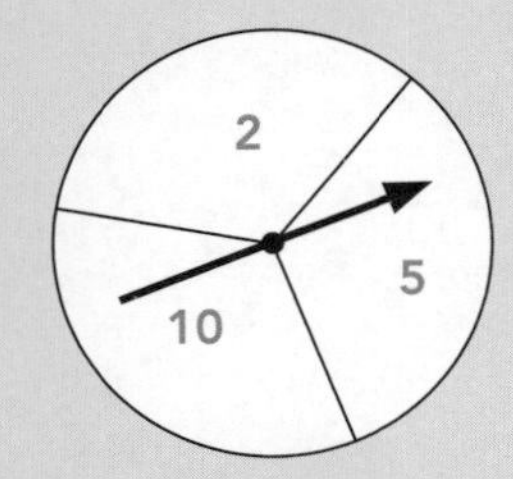

There are 3 equal sectors in this spinner, so the probability of landing on any of these sectors is $\frac{1}{3}$ or $0.\dot{3}$.

ISBN: 9780170477710

N

Calculating theoretical probability

- When we use devices such as coins, dice and spinners, we **know** the probabilities of single events.
- We can use these probabilities to calculate the probabilities of combined events.

The number of outcomes we are interested in.

$$\text{Probability} = \frac{\text{number of 'favourable' outcomes}}{\text{number of outcomes in the sample space}}$$

The total number of possible outcomes.

Example:
A marble is pulled from the bag shown without looking.

a What is the probability that it's a green marble?

$$\text{Probability} = \frac{\text{number of green marbles}}{\text{total number of marbles}}$$

$$\text{P(green)} = \frac{2}{5} \text{ or } 0.4$$

This means 'the probability of picking a green marble'.

b What is the probability that a marble is green or white?

$$\text{Probability} = \frac{\text{number of green and white marbles}}{\text{total number of marbles}}$$

$$\text{P(green or white)} = \frac{2 + 1}{5} = \frac{3}{5} \text{ or } 0.6$$

This means 'the probability of picking a green or white marble'.

c What is the probability that a marble is red?

$$\text{Probability} = \frac{\text{number of red marbles}}{\text{total number of marbles}}$$

$$\text{P(red)} = \frac{0}{5} \text{ or } 0$$

ISBN: 9780170477710

Answer the following.

1 Kai rolls a die.

a Calculate the probability of throwing a 3. P(3) = ________

b Calculate the probability of throwing a 1 or a 2. P(1 or 2) = ________

c Calculate the probability of throwing an odd number. P(odd) = ________

d Calculate the probability of throwing a 7. P(7) = ________

e Calculate the probability of throwing a 1, 2, 3, 4, 5 or 6. P(1, 2, 3, 4, 5 or 6) = ________

2 Calculate the following probabilities when using this spinner.

a P(green) = ________

b P(white) = ________

c P(grey) = ________

d P(white or green) = ________

e P(yellow) = ________

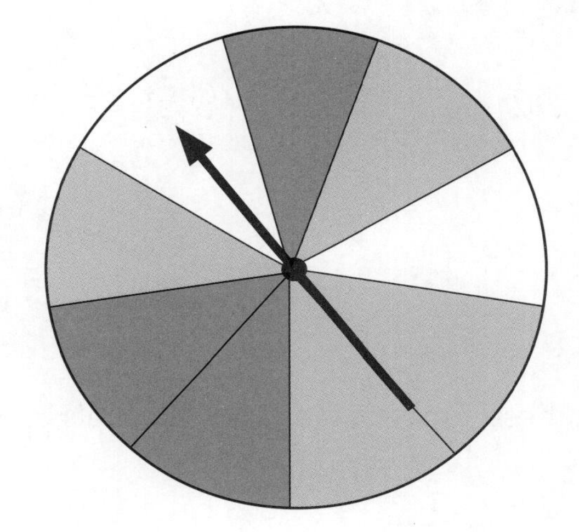

Calculate the probability of each event shown if **one** object is selected at random from this group.

3

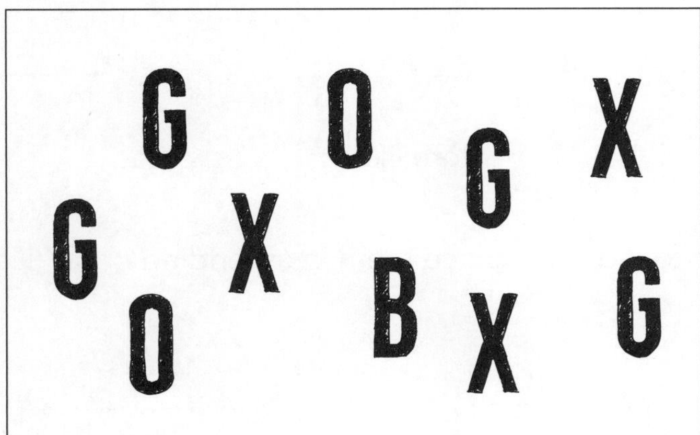

a P(O) =

b P(G or X) =

c P(O or B) =

ISBN: 9780170477710

Comparing probabilities

- Probabilities can be written as **fractions**, **decimals**, **percentages** or **proportions**.
- If you want to **compare** probabilities, it is usually easiest to convert them to **decimals**.
- You need to be able to convert probabilities to decimals using your calculator.

Examples: Convert these probabilities into decimals and state which is more likely.

Compare the hundredths columns.

1 $\frac{2}{5}$ or $\frac{5}{12}$

$\frac{2}{5} = 0.40$

$\frac{5}{12} = 0.41\dot{6}$

To change a fraction to a decimal:
2 [fraction key] 5 [=] [S⇔D]

1 is larger than 0, so a probability of $\frac{5}{12}$ is more likely than a probability of $\frac{2}{5}$.

÷100

2 $\frac{5}{6}$ or 79%

$\frac{5}{6} = 0.8\dot{3}$

79% = 0.79

Compare the tenths columns.

8 is larger than 7, so $\frac{5}{6}$ is more likely than 79%.

Convert these probabilities into decimals to 2 dp and state which is more likely.

1 $\frac{1}{5}$ = ________ 19% = ________

More likely: ____________

2 87% = ________ $\frac{11}{12}$ = ________

More likely: ____________

3 53% = ________ $\frac{4}{7}$ = ________

More likely: ____________

4 25% = ________ $\frac{2}{9}$ = ________

More likely: ____________

Convert these probabilities into decimals to 2 dp and place them in ascending order (least likely to most likely).

5 $\frac{3}{7}$ = ________ $\frac{6}{13}$ = ________ 41% = ________

____________ Least likely ____________ ____________ Most likely

 ISBN: 9780170477710

6 Which of these bags are you most likely to pick a green ball out of? Round the probabilities to 2 dp.

Bag A

P(green) = ________

Bag B

P(green) = ________

Bag C

P(green) = ________

Most likely is bag ________

7 Which of these spinners is most likely to land on green? Round the probabilities to 2 dp.

Spinner A

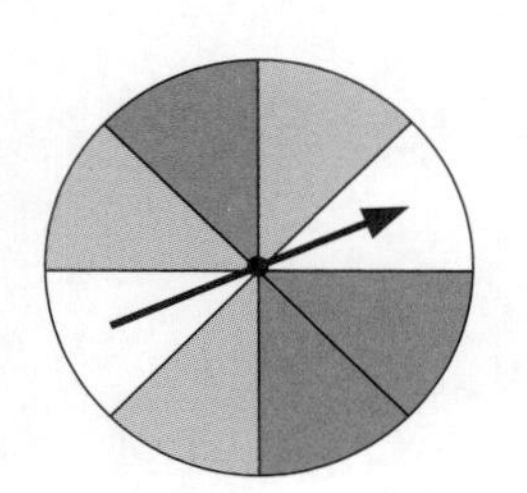

P(green) = ________

Spinner B

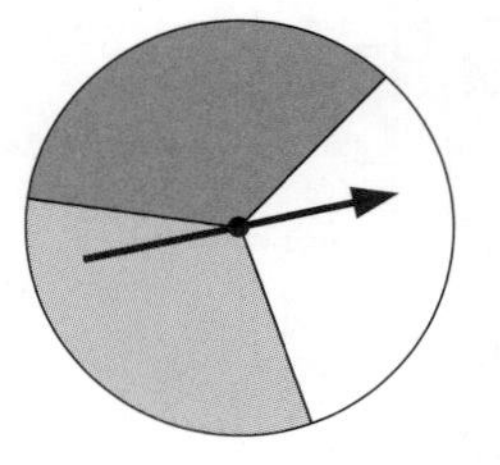

P(green) = ________

Spinner C

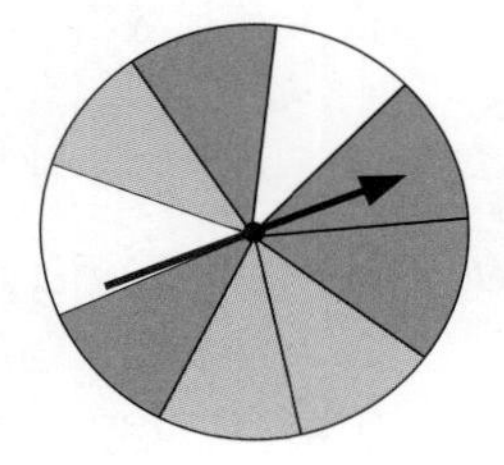

P(green) = ________

Most likely is spinner ________

8 Petra has a tetrahedral die. It is equally likely to land with a 1, 2, 3 or 4 at the top.

She claims that she is more likely to get a 4 at the top when she throws it, than she is to get two heads when she tosses two coins. Is she correct?

P(4 at top of die) = ________

P(two heads) = ________

Is she correct? Explain. ____

9 If you threw three normal dice together, the probability of getting three 6s = $\frac{1}{216}$.

If you toss seven coins at once, the probability that they all land on heads = 0.0078.

Which event is more likely? ________________________

10 If you bought a $7 lotto ticket, the probability of winning first division is $\frac{1}{383\,838}$.

If you tossed 18 coins at once, the chance of all 18 landing on heads is 0.0000038 (1 sf).

Which event is more likely?____

ISBN: 9780170477710

N

Complementary events

- **Complementary** events occur when there are **only two possible outcomes**, e.g. scoring a goal or not scoring a goal, rolling a die and getting an even number or an odd number.
- The probabilities of complementary events **add to 1**.

Notice this is spelt with an 'e'. Things that are complimentary are free.

Examples:

1 When tossing a fair coin, the result must be a head or a tail.

P(head) = 0.5
P(tail) = 0.5 } **0.5 + 0.5 = 1**

The events 'tossing a head' and 'tossing a tail' are complementary.

2 On a particular day, it either snows or it doesn't snow. So these events are complementary. The weather forecast says there is a 5% chance of snow.

The chance it will not snow is 100% – 5% = 95%.

Answer the following questions.

1 On Saturday, the weather forecast says that the probability of rain is 0.4.

The probability that it won't rain = __________.

2 The probability of rolling a 2 on a die = $\frac{1}{6}$ P(not throwing a 2) = ______________

3 In a family of nine children, three are blond. What is the probability that a child in this family is not blond?

P(not blond) = ______________

4 Twelve students in a class of 25 have at least one sister. Calculate the probability that a student doesn't have a sister.

P(student don't have a sister) = ______________

5 A school has a roll of 637. Five hundred and eighteen of these students are right handed. What is the probability of a student not being right handed?

P(student is not right handed) = ______________

6 Amira sells vegetable plants in the market. She planted 24 courgette seeds and only 18 germinated. What is the probability that one of her seeds did not germinate?

P(seed did not germinate) = ______________

ISBN: 9780170477710

Risk

- The word 'risk' means to the probability of something bad happening.

Example: In 2021, the number of people killed in car accidents was 319. The population was 5.123 million.

Therefore, the risk of being killed in a car accident in New Zealand in 2022 was:

$$\frac{319}{5.123 \text{ million}} = 0.000062$$

Answer the following questions.

1 On a particular day, these were the results of driver's licence testing.

Pass	24
Fail	13

On this day, what is the risk that you failed your licence test? ________

2 a In a board game, you lose your turn if you roll a 3 on a die.

What is the risk of losing your turn? ________

b If you land on a certain square on the board, you must spin the spinner.

What is the risk that you lose exactly $50? ________

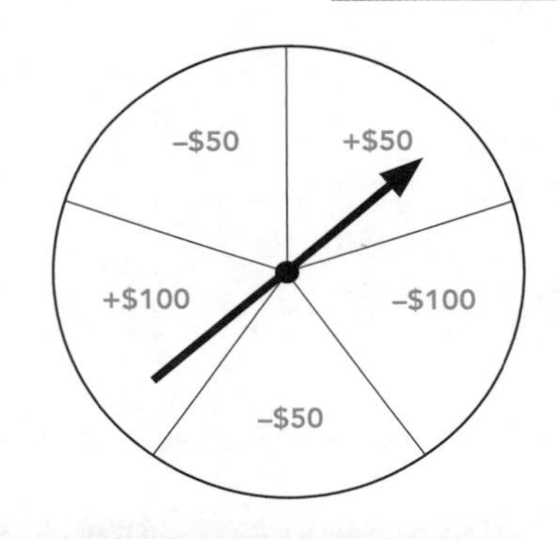

3 Kauri's lateness to classes has been recorded by his teachers.

	On time	Late	Total
Mathematics	45	9	54
English	40	8	48

a Based on this data, what is the risk that he will be late to the next Mathematics class? ________

b Based on this data, what is the risk that he will be late to the next Mathematics or English class? ________

4 Rupert has a bag containing 28 jelly beans. The white ones are vanilla but the green ones are vomit flavour. Without looking, he pulls out a jellybean.

What is the risk that it is vomit flavour? ________

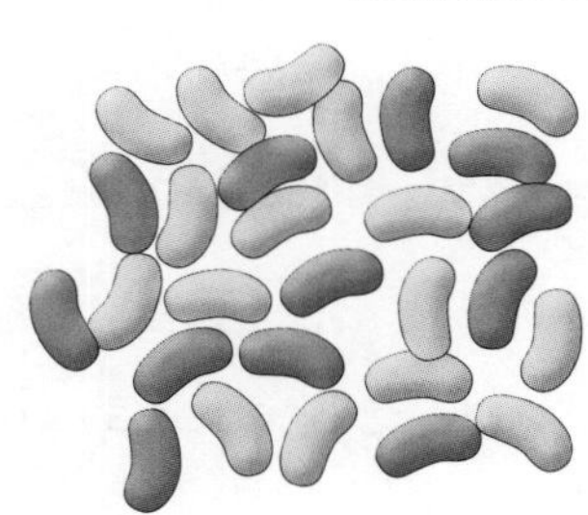

Process ideas

- In the Numeracy assessment you will need to be able to solve mathematical problems in a range of meaningful situations using three process ideas.

You will need to:
1 **Formulate** approaches to solving problems. This means working out how to solve a problem.
2 **Use** mathematics and statistics.
3 **Explain** whether answers and statements are reasonable.

1 Formulate approaches to solving problems

- These questions will require that you to work out **how** to solve the problem.
- Usually there are several steps needed to find the answer.

Example:
A Canada goose can fly 850 km in 8 hours. To the nearest minute, how long would it take to fly the 34 km from Timaru to Geraldine?

You could do this in two ways:

i Show the calculation with speech bubbles.

850 km takes 8 hours

So 34 km takes $8 \times \frac{34}{850} = 0.32$ hours.

Find the number of hours by multiplying by $\frac{34}{850}$.

0.32 hours = 0.32 x 60 minutes

= 19.2 minutes

= 19 minutes (0 dp)

Convert hours to minutes by multiplying by 60.

Round 19.2 to the nearest minute.

ii Explain each stage of the calculation in sentences.

If 850 km takes 8 hours, then 34 km will take $\frac{34}{850}$ of 8 km = 0.32 hours.

Convert 0.32 hours to minutes by multiplying by 60 → 19.2 minutes.
19.2 minutes to the nearest minute = 19 minutes.

Answer the following questions.

1 a Tina wants to go on an Outward Bound course which costs $5000. She has managed to save $1500 and her dad has offered to lend her the remaining $3500. She will pay him 5% interest each year for two years. She has worked out that she will need to repay him a total of $3850. Is she correct? Show your reasoning.

Yes/No. ____________________

____

b She repays him the same amount each week for two years.
To the nearest dollar, how much will each weekly repayment be? $________

 ISBN: 9780170477710

2 A kiwi egg has a mass of 372 g of which 60% is yolk. A hen egg on average has a mass of 60 g and 35% is yolk. Alex calculated that he would need 11 hen eggs to get the same amount of yolk that is contained in one kiwi egg. Is he correct? Explain how he could have reached this conclusion.

____

____

3 Liam needs four herb plants. Single plants normally cost $6.90. In its sale, the shop has two deals.

Deal 1: For each plant bought at full price, you can buy a second plant at half price.

Deal 2: Buy plants at 20% cheaper than the normal price.

Thyme Parsley Mint Oregano

Which deal is better? Explain your answer, and show your calculations.

____

____

4 Archie took ¾ hour to walk 3.6 km. At this rate, how long would it take him to walk 10 km? Give your answer in hours and minutes.

____

____

5 The graph shows the daily pay rates for picking punnets of raspberries for two growers, Royal Raspberries and Ruby Raspberries.

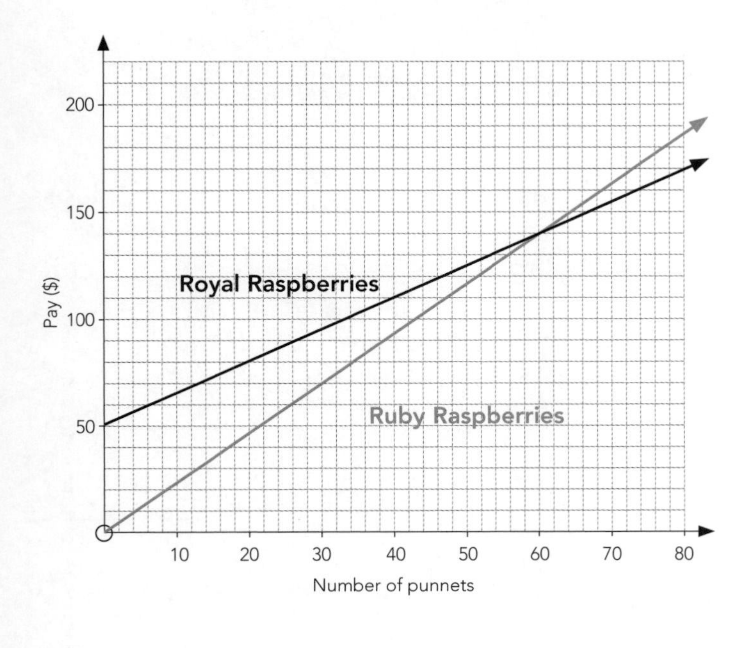

a Complete this statement:

Royal Raspberries pays $________ per day plus $________ for each punnet picked.

b Arlo is confident that he can pick 80 punnets in a day. Which grower should he work for, and how much more would he earn each day?

Grower: ______________________

____

Difference in pay: ______________

ISBN: 9780170477710

N

2 Use mathematics and statistics

- In these type of questions you will be **told what** to calculate.

Example:
Abbie's boyfriend is in London. New Zealand time is 13 hours ahead of London time. She wants to ring him at 19:45 London time. What time will that be in New Zealand?

New Zealand is 13 hours ahead, so add 13 hours to the London time:

$$\begin{array}{r} 19{:}45 \\ +\ 13{:}00 \\ \hline 32{:}45 = 24 + 08{:}45 \end{array}$$

Answer the following questions.

1 Hamish is making bunting to decorate his house. The bunting is made up of small triangular flags sewn onto a length of cord.

He cuts the triangles from strips of fabric that are 24 cm deep and 160 cm wide.

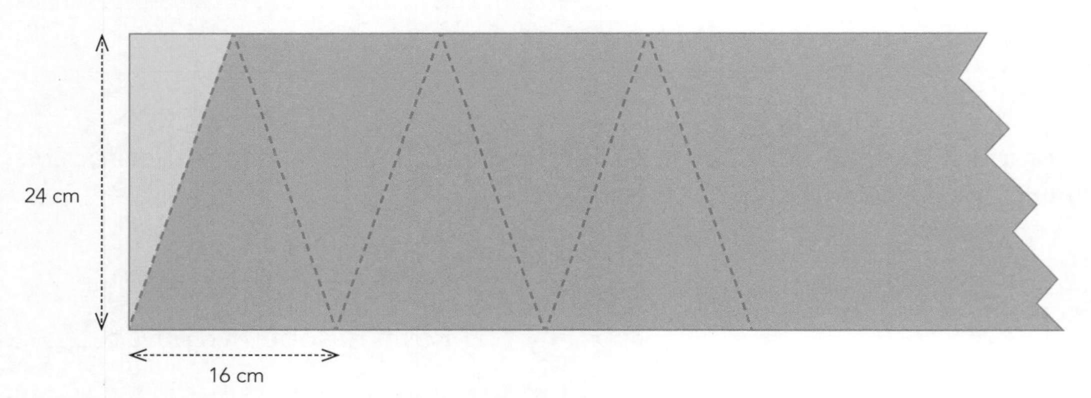

a How many flags can he get from one strip of fabric? ________ flags

b Calculate the area of fabric needed for each flag. ________ cm^2

c The flags will be spaced on the cord so that their centres are 18 cm apart. He has calculated that he needs 15.3 m of bunting. How many flags will he need to make? ________ flags

d He will need a length of 1.5 m of fabric which costs $11.56 per metre. How much will it cost him? $________

e The normal price for a different fabric is $14.80 per metre. However, it is on special, so the normal price is reduced by 35%. Calculate the reduced price per metre. $________

ISBN: 9780170477710

2 Melanie collected and measured some turret shells from a beach.

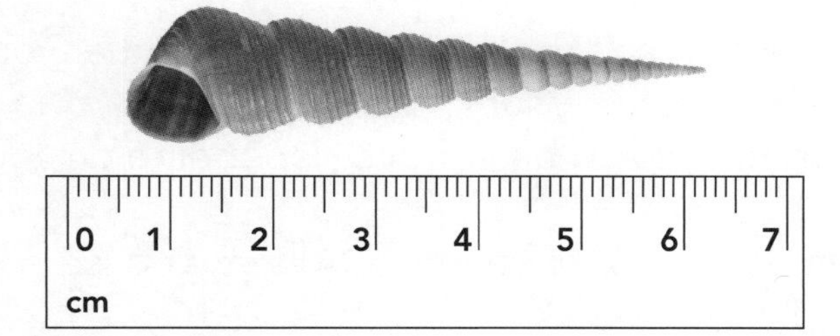

a How long is this turret shell? Length = ________ cm

b She displayed their lengths on a graph. What type of graph is this?

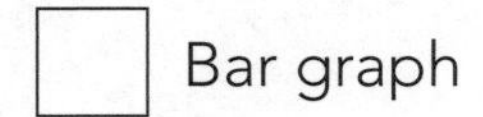

☐ Bar graph

☐ Histogram

☐ Strip graph

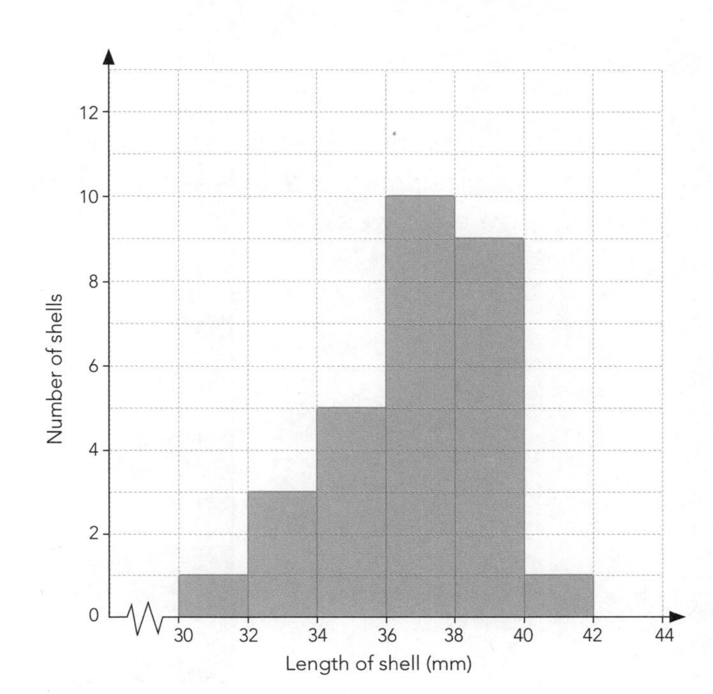

c How many turret shells did she measure and graph?

____

d She found one more shell that was exactly 36 mm long. Add this value to the graph.

3 Toni is threading beads to make a necklace. She needs:

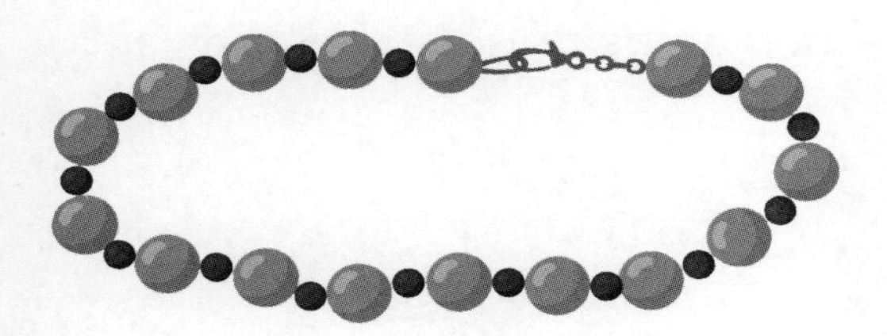

16 large beads at 80 cents each,

15 small beads at 45 cents each,

and a piece of chain with a catch, which costs $1.75.

Calculate the total cost of the beads, the chain and the catch. Cost = $________

4 Moana makes wooden puzzles. This figure is made from four cubes stuck together. She sits this on her workbench and spraypaints it. How many square faces will be painted?

________ faces

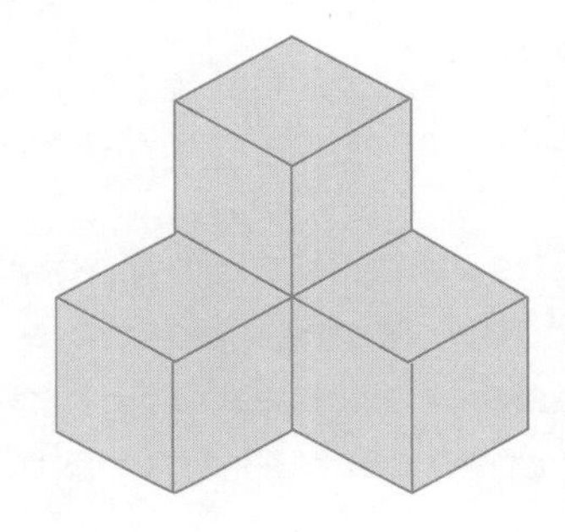

N

3 Explain the reasonableness of responses

- In these questions you will be given a statement and you need to state whether you agree, disagree or can't tell for sure. Then you **must** explain your choice.
- Sometimes more than one of the three options could be correct, depending on your explanation.
- **Do not** write answers based on your own opinion or experience. You **must** base your answer on **information given to you in the question**.
- Where possible, **use numbers** to support your answer.
- When answering questions about calculations, **use words** in your explanations.

Example:

The graph shows on average the results from a survey of Year 9, 11 and 13 students. They were asked how much time last week they spent active during PE classes, training or playing organised sport each week.

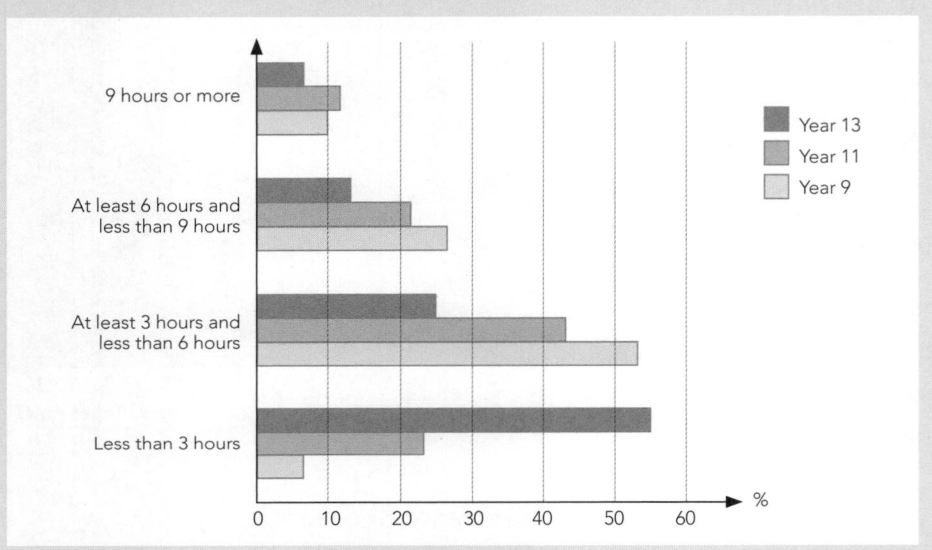

A newspaper headline says 'Declining student activity — most students doing less than an hour of exercise each day'.

☑ Agree ☐ Disagree ☐ Can't tell for sure

From the graph, about 80% of Year 13 students, 66% of Year 11 and 60% of Year 9 do less than 6 hours of the listed activities each week.

Where possible, use numbers to strengthen your argument.

or ☐ Agree ☐ Disagree ☑ Can't tell for sure

The data is from one week only, so without knowing how much activity is generally done, it's not possible to say whether student activity is declining. Also, the survey asked only about PE classes, training or playing sport. There are other ways of being active such as tramping.

ISBN: 9780170477710

Answer the following questions.

1 Elsie sells jars of jam, jelly, chutney, pickle and lemon curd at the local market. The pie graph show the percentages of each type sold during the last year.

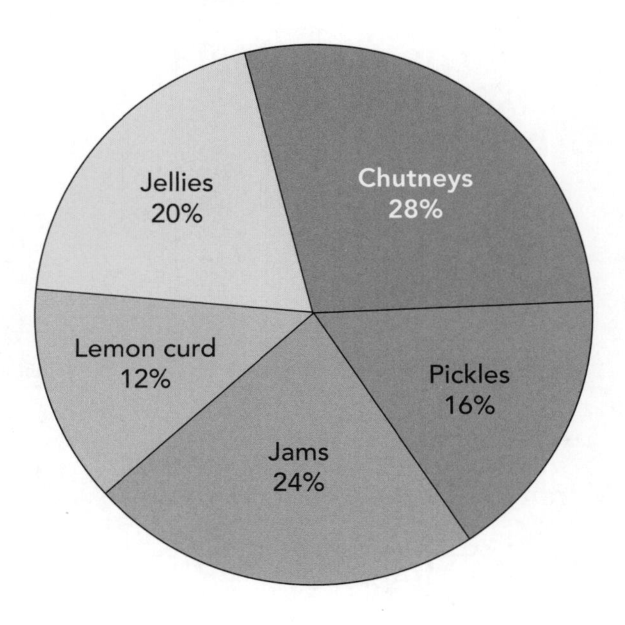

She made a profit of $2 on each jar of pickle and chutney and $1 profit on each jar of jelly, lemon curd and jam. Her friend calculated that she made 61% of her profit on pickles and chutney.

☐ Agree ☐ Disagree ☐ Can't tell for sure

Explain your answer. ____________________

____

2 Miru is investigating prices for next year's firewood. She has three possibilities.

A Frankie's Firewood: 3.6 cubic metres for $400, no delivery charge.

B Wallace's Wood: 7.2 cubic metres at $95 per m^3 + $36 for delivery.

C Fergie's Fuel: $108 for 1.2 cubic metres, which she can collect from nearby using her neighbour's trailer.

Which option gives her the lowest price per cubic metre? Option ______

Explain your answer. ____________________

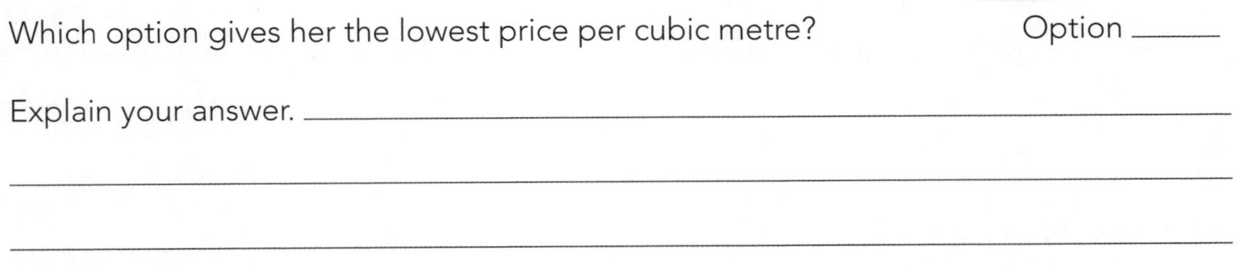

ISBN: 9780170477710

N

3 The school librarian produced this graph to show the numbers of books borrowed during the year.

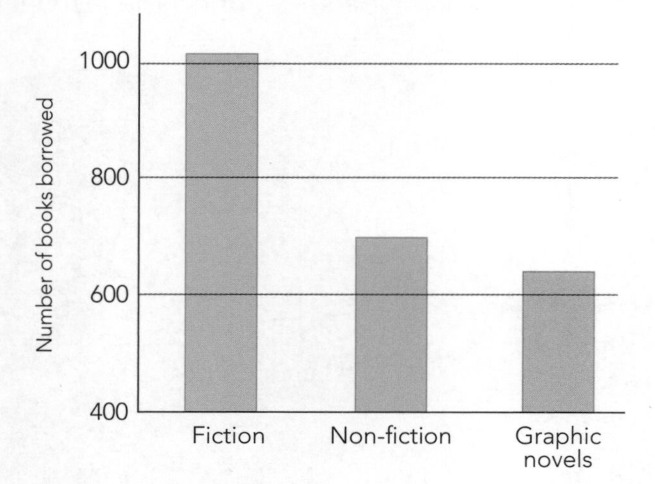

More than twice as many fiction books than non-fiction books were borrowed.

☐ Agree ☐ Disagree ☐ Can't tell for sure

Explain your answer. ____

____

4 This is the sign at a stall at the school fair:

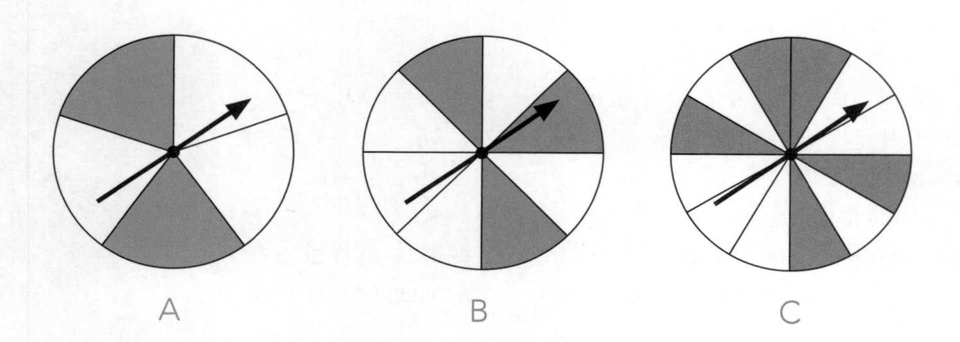

Eloise says that spinner A has the best chance of winning and spinner C has the worst chance.

☐ Agree ☐ Disagree ☐ Can't tell for sure

Explain your answer. ____

____

ISBN: 9780170477710

5 Cheryl's dad grows apples. To be suitable for export, apples need to have a mass of at least 160 g. Cheryl takes a sample of 380 apples and records the mass of each one. The graph shows her results. Her dad's target is to have 50% of his apples heavy enough for export.

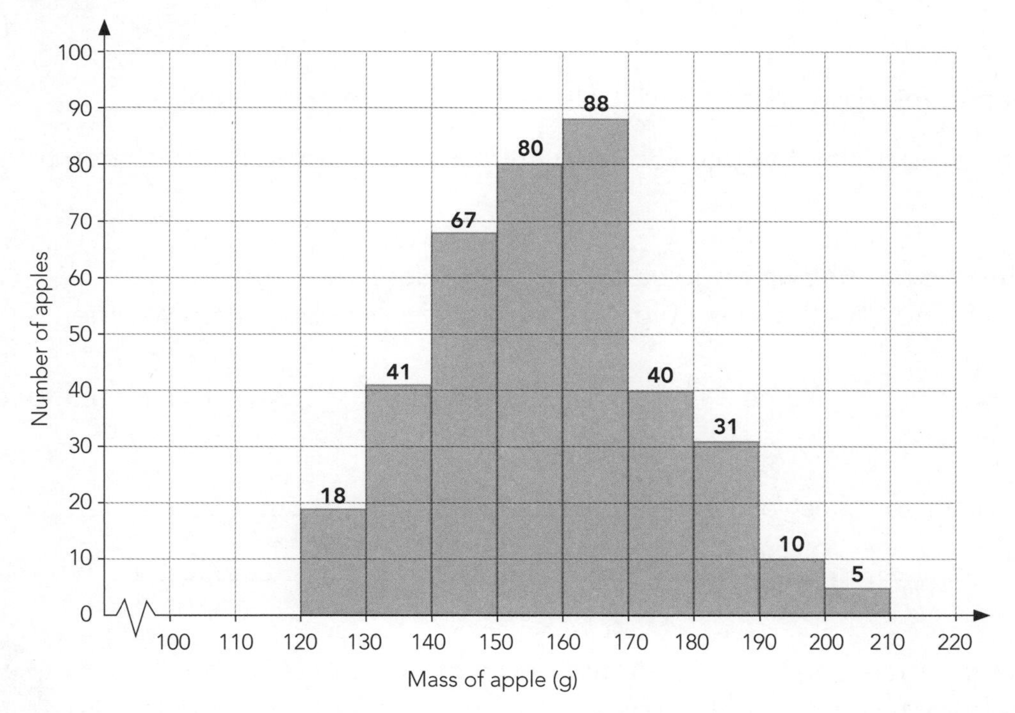

Does this sample suggest that he will meet his target? Yes/No

Explain your answer. ______________________________

____

6 In New Zealand, we use degrees Celsius (°C) to measure temperature. Some other countries use degrees Fahrenheit (°F).

To convert °F to °C, you need to subtract 32 and then divide by 1.8.

Robert looks at this thermometer and decides that 100°F is the same as 37°C. Is he correct?

Yes/No

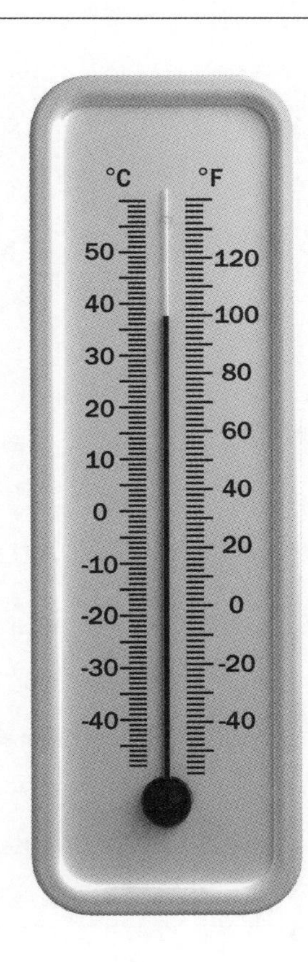

Explain your answer. ______________________________

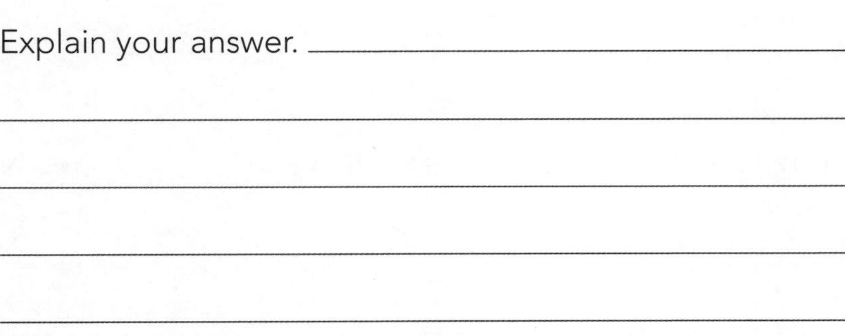

N

7 George is playing Monopoly with a friend.
The game is played using two dice.

Sometimes players throw 'doubles' — two of the same number.

The probability that this occurs = $\frac{1}{6}$. If a player throws doubles, then they get another turn.

George has thrown doubles on each of his last three turns.
His friend says that he couldn't possibly throw doubles a fourth time.

☐ Agree ☐ Disagree ☐ Can't tell for sure

Explain your answer. ______________________________

____

8 A ruru (morepork) is sitting at the origin of these axes, with its body facing down the *y*-axis. It can turn its head horizontally through an angle of 270°.
A weta (desirable food) sits on a nearby branch at the point (6, 4).

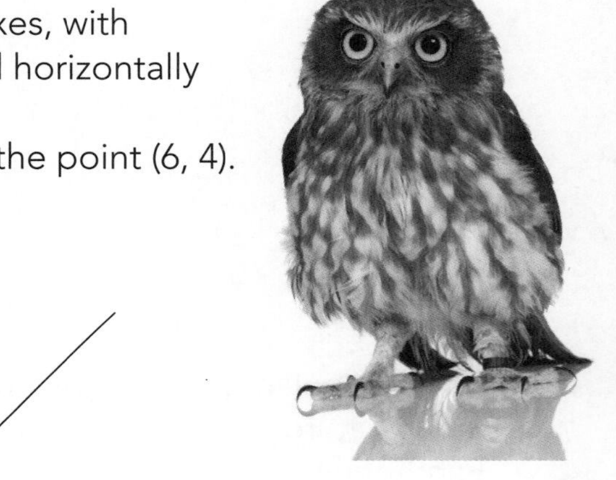

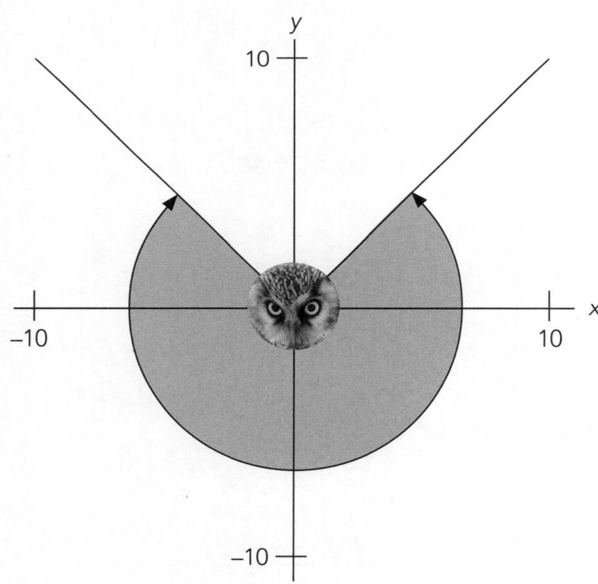

Will the ruru be able to turn its head far enough to see the weta? Yes/No

Explain your answer. ______________________________

____

 ISBN: 9780170477710

9 a Kate runs a coffee cart. One day she counted the number of cups of coffee ordered by each customer. There were 92 orders in total. She drew a graph of the results.

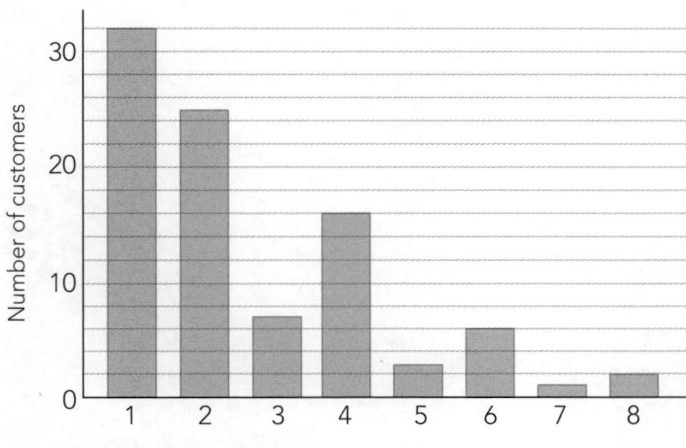

She is considering providing holders for carrying four cups of coffee at once. Her accountant has said that it will be worth her while only if 25% or more of her customers order at least four cups at once.

Does her data for this day support buying cup holders? Yes/No

Explain your answer. ______________________________

____

b Her weekly profits depend on how many cups of coffee she sells. She has to pay interest on the money she borrowed in order to buy the coffee cart, rent for the space it occupies, and she has to buy the cups and ingredients she uses.

The graph shows the relationship between the number of cups of coffee sold and her profit each week.

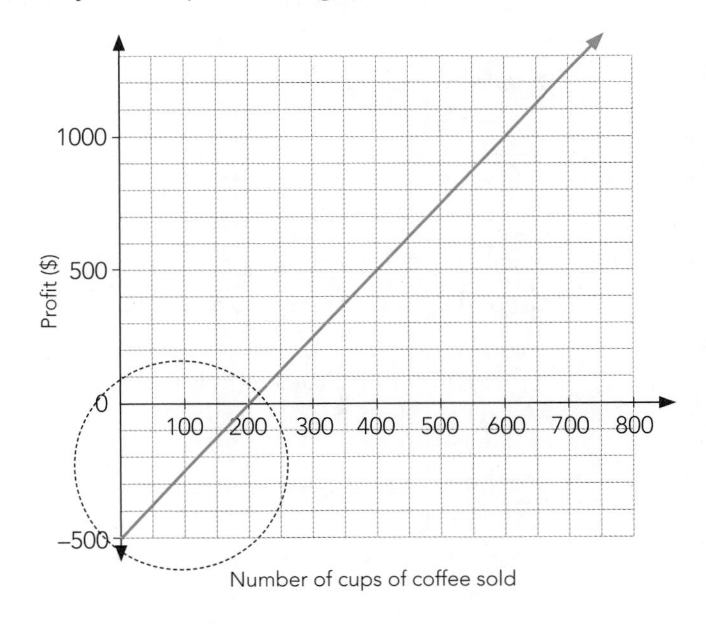

Explain the meaning of the circled part of the graph.

____

____

ISBN: 9780170477710

walkermaths

Numeracy practice

Set 1

1 **a** The pie graph shows the average proportions of animals that make up the diet of harrier hawks.

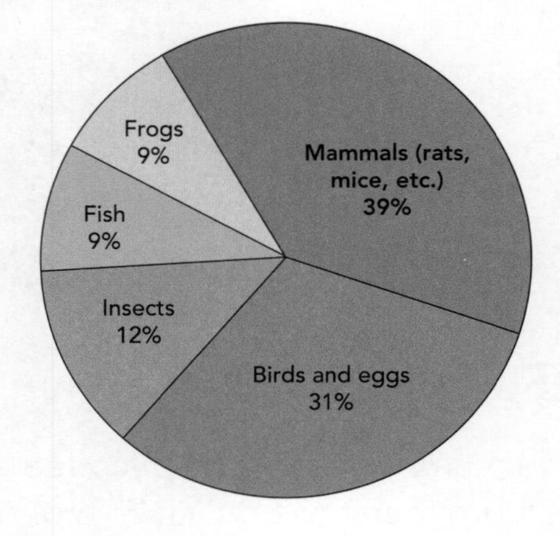

If a harrier hawk ate 5 kg of food over a period, how many grams of insects would you expect it to have eaten?

_____________ g

b The food eaten by a harrier hawk contained 124 g of birds and eggs. Estimate the total mass of food consumed during this time.

_____________ g

c Troy looked at the graph and said that a third of the diet of a harrier hawk was made up of frogs, fish and insects.

☐ Agree ☐ Disagree ☐ Can't tell for sure

Explain your answer. ____

____

____

2 **a** Sophie bought a new car. Her old one used 14 L of fuel per 100 km. She can drive 12.8 km in her new car for every litre of fuel used.

Is her new car more efficient than her old one? Yes/No

Explain your answer. ____

____

____

b The car has a height of 1595 mm. The height of the opening into her garage is 1.715 m. How big will the gap be between the top of her new car and the top of the opening?

_____________ mm

ISBN: 9780170477710

3 a Cindy collected information on the method of transport students used to get to school. Tick ALL the true statements.

Transport	Tally
Car	𝍸
Bus	||||
Walk	𝍸 |||
Bike	𝍸 |
Scooter	|||

☐ More students travel to school by bus than by car.

☐ The most common way of getting to school is by walking.

☐ Students are twice as likely to bike as they are to scooter to school.

☐ Most students get to school using something with wheels.

b Which of these graphs shows the data correctly? There may be more than one answer.

☐ 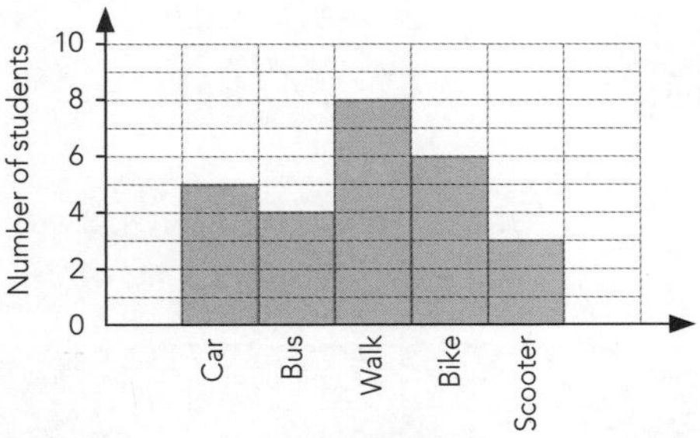

☐ 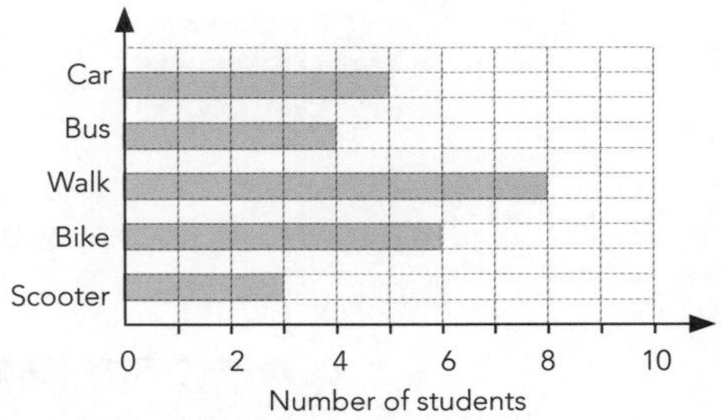

☐ 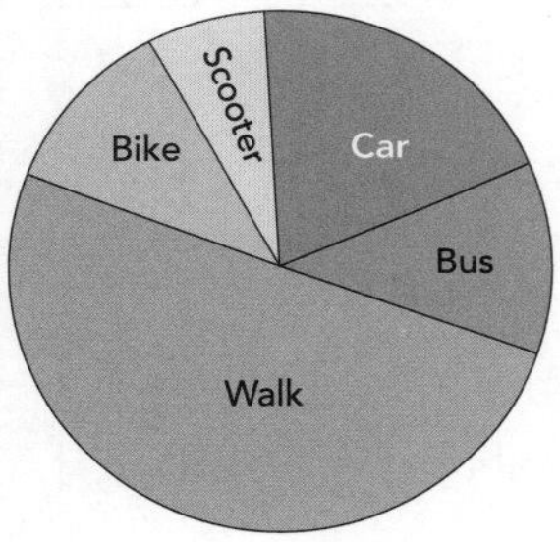

☐ 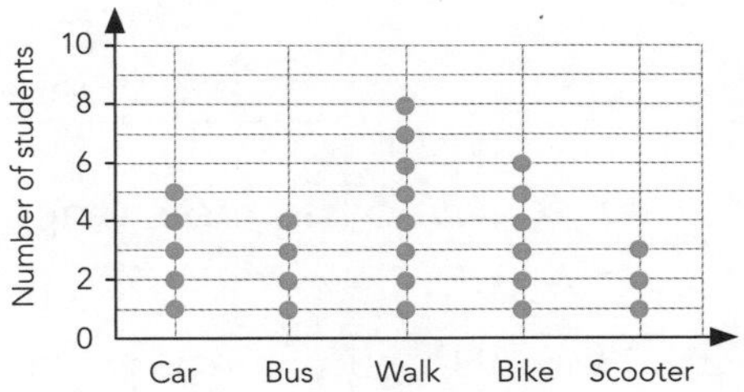

c Louisa scooters to school. All 14 bars light up when her battery is fully charged. What percentage of charge is left in this battery?

☐ 65–70% ☐ 70–75% ☐ 75–80% ☐ 80–85%

d On a full charge, the scooter can travel 50 km. If her house is 5.3 km away from school, how many days can she get to school and back without charging her scooter?

☐ 4 ☐ 5 ☐ 9 ☐ 10

e This is the logo on her scooter.
How many lines of symmetry does it have? ___________

ISBN: 9780170477710

N

Set 2

1 **a** Henrietta's phone is face down on a table. James is sitting opposite her.

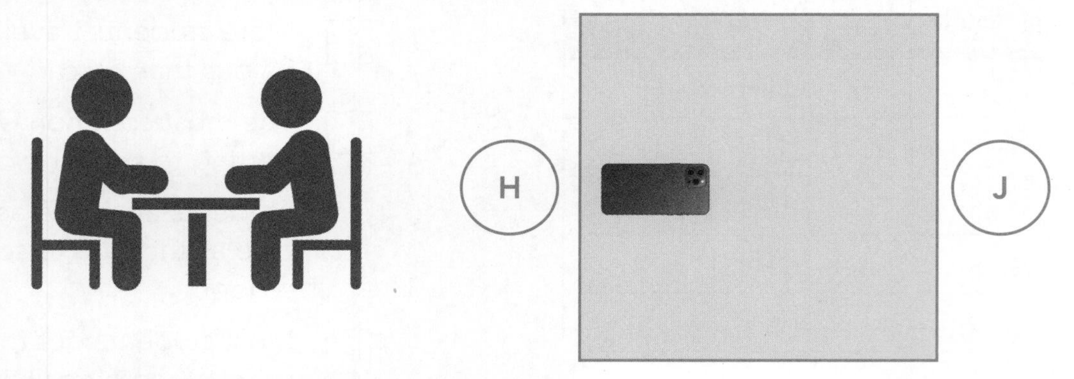

She rotates her phone 90° clockwise. Which view of Henrietta's phone would James see?

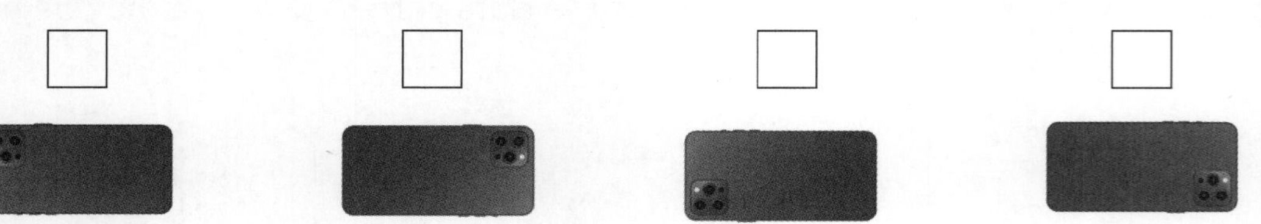

b James wants to buy a new phone like Henrietta's. Two shops have the model he wants in store.

Phoney	Celly
Normal price $799.99 This week's special: 15% off	Normal price $749.00 This month's deal: 10% off

Who has the cheapest phone and how much will it cost? ______________________

2 **a** Jeremy needs some Thai baht for his holiday in Thailand.

1 New Zealand dollar buys 21 Thai baht (THB)

How many New Zealand dollars will he need for 12 600 Thai baht? ______________

b Durian is a stinky fruit that is available in Thailand.

At a stall they were charging 180 Thai baht per kg.

What is the cost of a durian of mass 1250 g?

____________ Thai baht

c Thailand is six hours behind New Zealand time, e.g. when it is 11 pm in New Zealand, it is 5 pm in Thailand.

Jeremy left Auckland at 8:00 am. The flight to Thailand takes 11 hours and 50 minutes.

What will be the time in Thailand when he lands? ____________ am/pm

 ISBN: 9780170477710

3 **a** Henry and his family travelled from Auckland to Great Barrier Island on a ferry. In which direction did they travel?

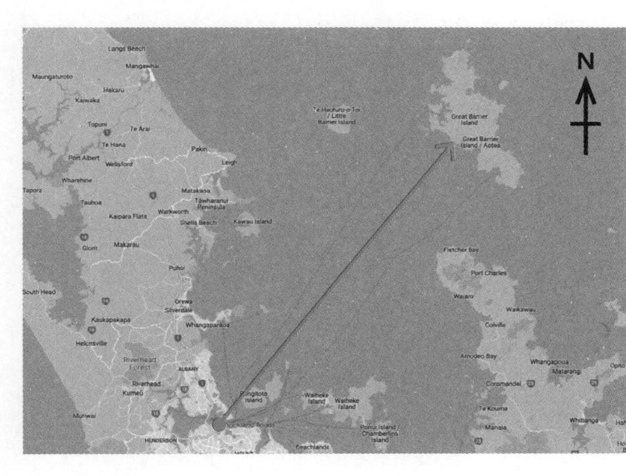

- ☐ SW
- ☐ NE
- ☐ NW
- ☐ SE

b The one-way ferry prices are in the table below.

Type	Fare
Adult	$105.50
Child (< 16 years)	$77.50
Student	$90.50

Calculate the total cost of ferry tickets for two adults, two children and one student.

$__________

c The ferry left at 8:02 am and it takes four and a half hours to get to Great Barrier Island. At what time did they arrive on the island?

__________ am/pm

d The distance the ferry covers during the four-and-a-half-hour trip between Auckland and Great Barrier Island is 99 km. Calculate the average speed of the ferry.

__________ km/h

e The local newspaper headline was 'At least half local ferry arrivals are late'.

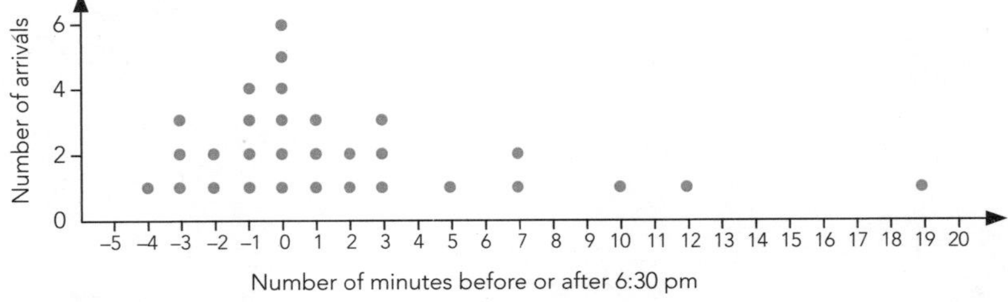

☐ Agree ☐ Disagree ☐ Can't tell for sure

Explain your answer. ______________________________

____

ISBN: 9780170477710

N

Set 3

1 **a** The table below shows the masses and prices of mandarins at the local supermarket.
Plot the three values on the graph below. Draw a line through the points and continue the lines to the edges of the graph in each direction.

Mass	Price
500 g	$3.25
1 kg	$6.50
2 kg	$13.00

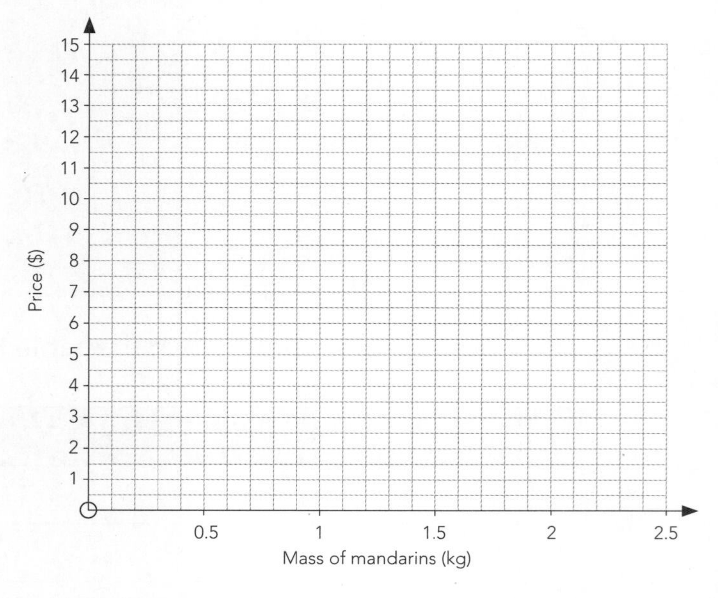

b Use the graph to estimate the mass of mandarins that could be bought for $10

_____________ kg

c If the supermarket increased the price of mandarins,

☐ the y intercept would increase.

☐ the y intercept would decrease.

☐ the gradient of the line would become steeper.

☐ the gradient of the line would become less steep.

d Esther wrote each of the letters for the word 'MANDARINS' on small squares of paper and placed them in a bowl. She closed her eyes and pulled out a paper square. What is the probability that the square had an A or an N on it?

P(A or N) = _____________

2 Which of the nets below could create this cube?

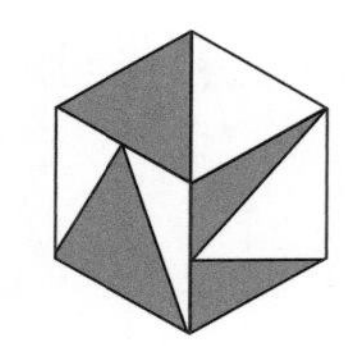

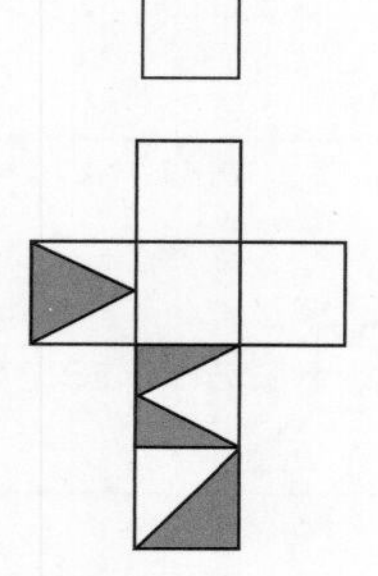

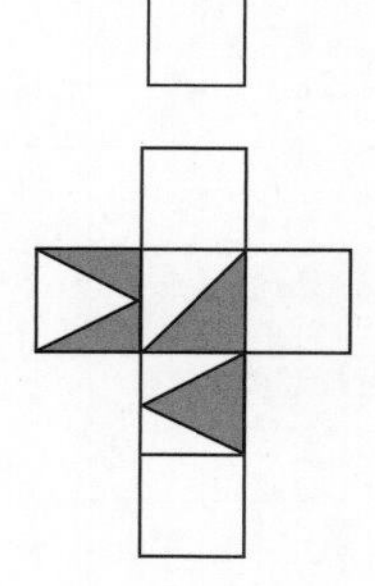

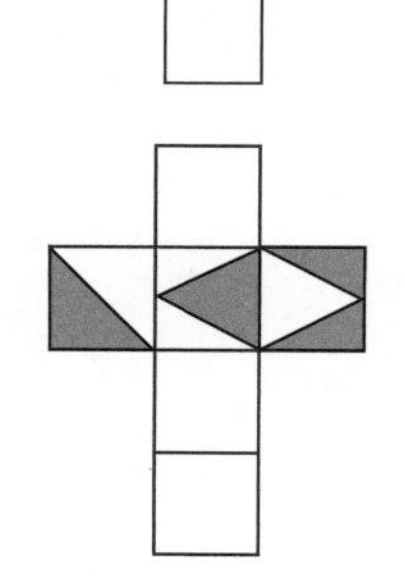

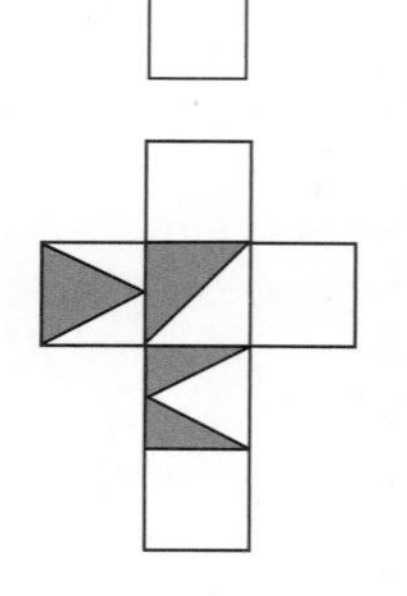

 ISBN: 9780170477710

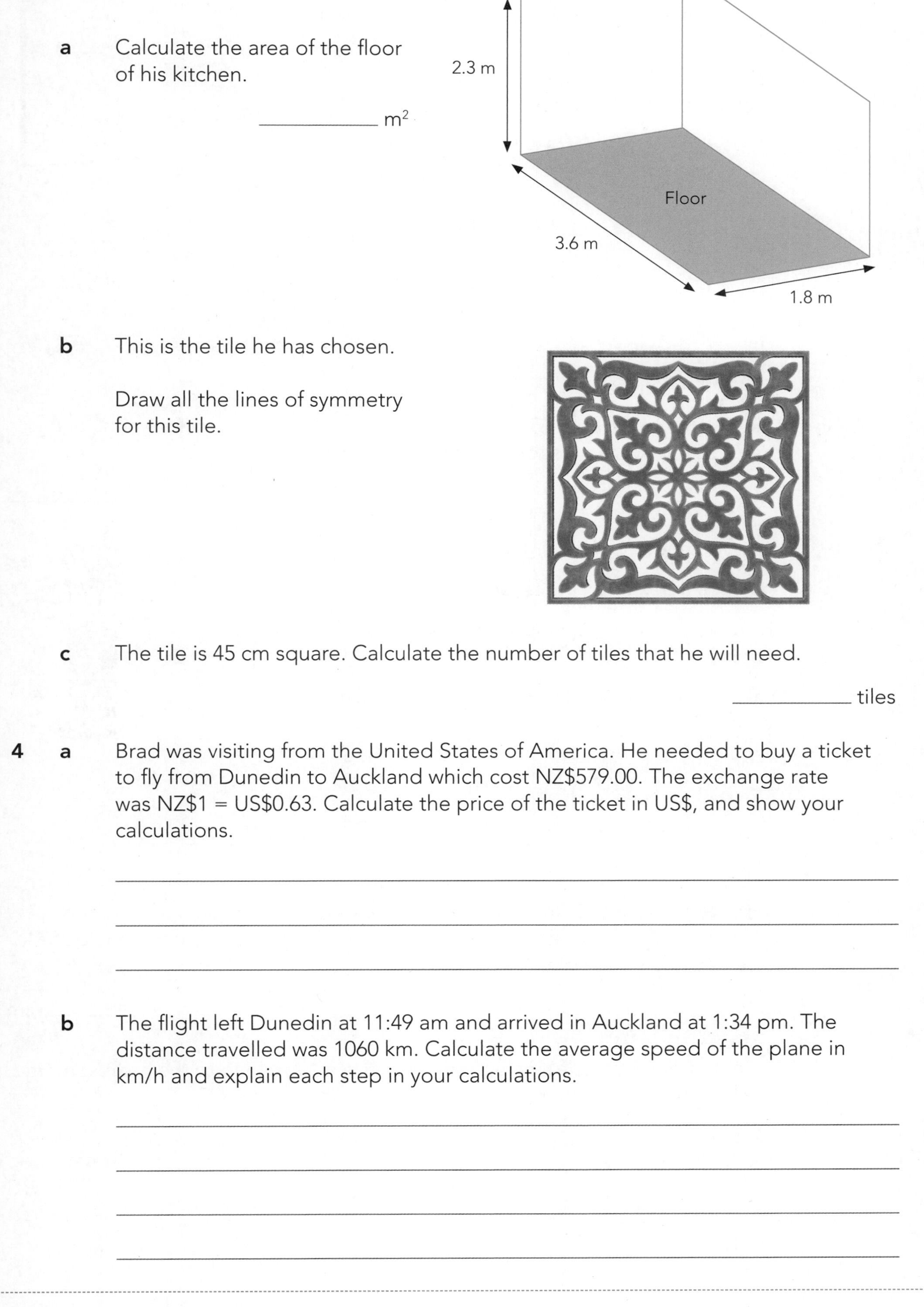

3 Vili wants to tile the floor of his kitchen.

a Calculate the area of the floor of his kitchen.

_________ m^2

b This is the tile he has chosen.

Draw all the lines of symmetry for this tile.

c The tile is 45 cm square. Calculate the number of tiles that he will need.

_________ tiles

4 a Brad was visiting from the United States of America. He needed to buy a ticket to fly from Dunedin to Auckland which cost NZ$579.00. The exchange rate was NZ$1 = US$0.63. Calculate the price of the ticket in US$, and show your calculations.

____

____

____

b The flight left Dunedin at 11:49 am and arrived in Auckland at 1:34 pm. The distance travelled was 1060 km. Calculate the average speed of the plane in km/h and explain each step in your calculations.

____

____

____

____

ISBN: 9780170477710

N

Set 4

1 Luka is getting this tattoo on his 27th birthday.

a When he holds his arm up in the mirror, which reflection will he see?

☐ ☐ ☐ ☐

b The cost of the tattoo is $750. Luka's mum has said that she will pay 40% of the cost. If Luka pays the rest of the cost, how much will he need to pay?

$________

2 a Gertrude needs 12 solar lights for her drive.

She finds a box of six lights for $49.95.

The store is offering a 20% discount if she buys two boxes of lights.

Which calculation would enable her to calculate the cost per light?

☐ 2 x 49.95 x 0.8 ÷ 12 ☐ 2 x 49.95 ÷ 0.8 ÷ 12

☐ 2 x 49.95 x 0.2 ÷ 12 ☐ 2 x 49.95 ÷ 0.2 ÷ 12

b Gertrude's drive is 18.7 metres long. She wants a light at each end, with the remaining 10 lights at equal intervals. How many centimetres apart should she place her 12 garden lights?

________ cm

c The path from Gertrude's garage to her house is 12 m long to the nearest metre. Which of these could be its length to the nearest centimetre?

☐ 1149 cm ☐ 1151 cm ☐ 1250 cm ☐ 1251 cm

ISBN: 9780170477710

3 Here is part of the timetable for Angelo's local bus.

Grey St	Maple Grove	Tawa Rd	Oak St	Rimu Lane	Totara Way	Daisy Lane	Elizabeth St
7:40 am	7:41 am	7:44 am	7:48 am	7:50 am	7:59 am	8:00 am	8:05 am
7:50 am	7:51 am	7:54 am	7:58 am	8:00 am	8:09 am	8:10 am	8:15 am
8:00 am	8:01 am	8:04 am	8:08 am	8:10 am	8:19 am	8:20 am	8:25 am
8:10 am	8:11 am	8:14 am	8:18 am	8:20 am	8:30 am	8:31 am	8:37 am
8:20 am	8:21 am	8:24 am	8:28 am	8:30 am	8:40 am	8:41 am	8:47 am
8:30 am	8:31 am	8:34 am	8:38 am	8:40 am	8:50 am	8:51 am	8:57 am
8:40 am	8:41 am	8:44 am	8:48 am	8:50 am	9:00 am	9:01 am	9:07 am
8:50 am	8:51 am	8:54 am	8:58 am	9:00 am	9:10 am	9:11 am	9:17 am

a Angelo needs to catch a bus so that he arrives at work by 9:00 am. His work is a 10-minute walk from Daisy Lane. What bus should he catch from Grey Street?

_____________ am

b A bus leaves Tawa Road at 8:44 am. How long does it take to reach Elizabeth Street?

_____________ min

c The times for the trip between Grey Street and Elizabeth Street are the same for the first three buses. The last five buses take a little longer to do the same route. Which sections of the trip take longer? Suggest a reason for this.

____

____

4 **a** Chloe has taken two trailers of green waste to the dump. The first contained 3 m^3 of green waste and the second contained 5 m^3. It costs $22 to dump a trailer of green waste, regardless of how much it contains. How much did it cost her on average for each cubic metre of green waste?

$_____________

b The dump is 11 km from her home. If it took her 12 minutes to drive there, what was her average speed in km/h?

_____________ km/h

N

Set 5

1 **a** You have 206 bones in your body, 52 of them in your feet. What percentage (to 2 dp) of your bones are in your feet?

__________%

b When human bones are found, the height of their owner can be calculated from the length, in centimetres, of the thigh bone (femur). For females, multiply by 2.47 and add 65.53. Calculate the height (to the nearest centimetre) of a woman whose thigh bone length is 40 cm.

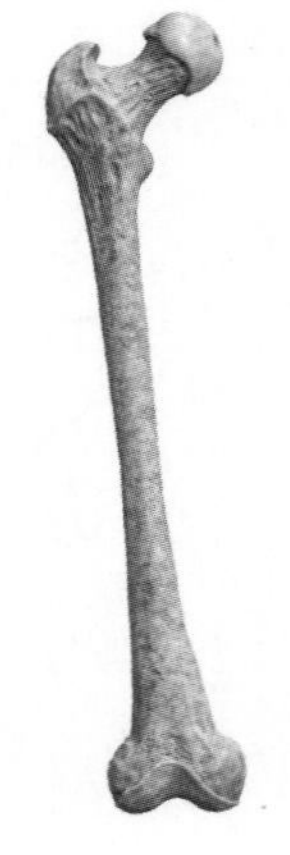

__________ cm

c The thigh bone measurement had been rounded to 40 cm. Which of these could be its actual length?

☐ 39.50 cm ☐ 39.05 cm ☐ 40.50 cm ☐ 40.51 cm

d The graph shows femur lengths from a sample of 150 New Zealand women.

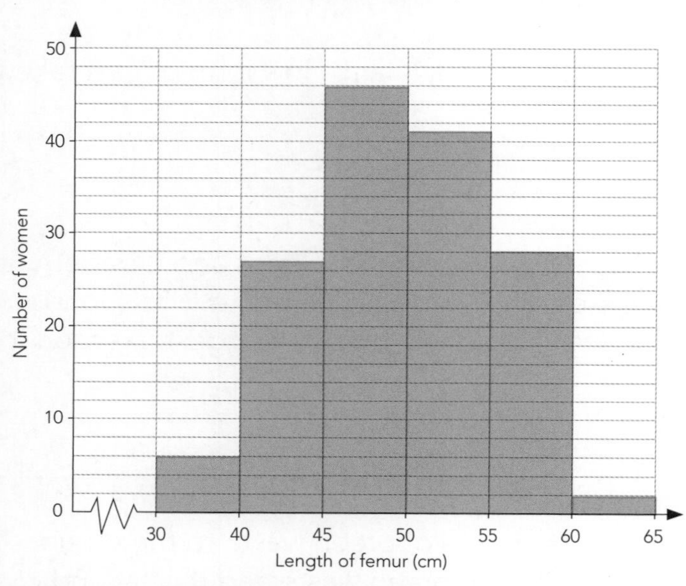

A report on this data said 'Most women have femur lengths between 45 and 55 cm'.

☐ Agree ☐ Disagree ☐ Can't tell for sure

Explain your answer. ____

____

e One extra value needs to be added to the graph. This thigh bone is 40 cm long. Put a cross (X) in the column to which this piece of data should be added.

 ISBN: 9780170477710

2 a Which of these two block structures could be fitted together to make a perfect $2 \times 2 \times 2$ cube?

☐ ☐ ☐ ☐

b How many faces of the cubes are visible on the shaded figure? ________ faces

3 a Putrid Petting Zoo prices to visit are:

- Adults $24
- Children $13
- Students or seniors $20
- Family pass (2 adults and 2 children) $60.00

ZOO

There is a sign outside which states:

SAVE $15! BUY A FAMILY PASS!

Is the advertisement true? Explain your answer using the information above.

____

____

____

b Last year, Putrid Petting Zoo had 12 660 visitors; 30% of them were international tourists.

How many visitors to the zoo were international tourists? ________

c The zoo has a display of the shells of the large carnivorous snails that live locally.

0 1 2 3 4 5 6 7 8 9 10 11
cm

Estimate the width of this snail shell. ________ cm

Answers

Operations on numbers (pp. 4–43)

Place value (pp. 4–6)

1

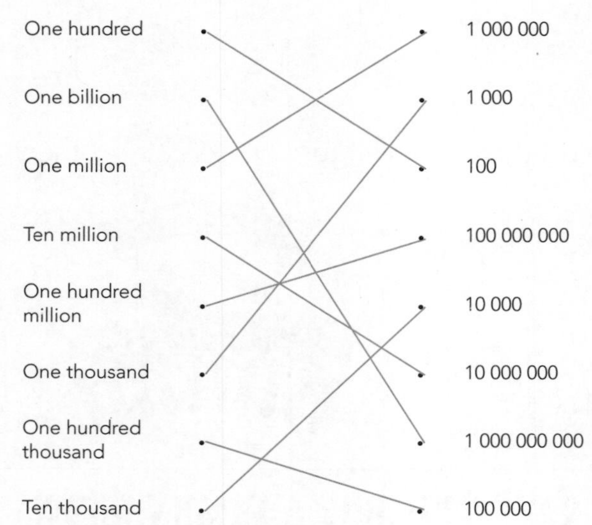

		Numerals	Words
2	23 491	400	Four hundred
3	1 272 408	70 000	Seventy thousand
4	7 419 364 201	7 000 000 000	Seven billion
5	442 196 939	2 000 000	Two million
6	675 521	600 000	Six hundred thousand
7	913 841 725	900 000 000	Nine hundred million
8	431 612	1 000	One thousand
9	182 341 092	80 000 000	Eighty million

10

Two thousand, five hundred and eighty-four	Two million, five hundred thousand, eight hundred and four	Two billion, five thousand, eight hundred and forty
2 584	2 500 804	2 000 005 840
Two million, fifty thousand, eight hundred and forty	Twenty thousand, five hundred and eighty-four	Two hundred and five thousand, eight hundred and forty
2 050 840	20 584	205 840
Two and a half billion, eight hundred thousand and forty	Two million, five thousand, eight hundred and forty	Two hundred and fifty thousand and eighty-four
2 500 800 040	2 005 840	250 084

11 Eighty two thousand, one hundred and thirty-seven

12 Eight million, ten thousand and three hundred

13 Two hundred and four thousand and nine

14 Two billion, one million and thirty-two thousand

15 Nine hundred and one million, fourteen thousand, five hundred and sixty-one

16 64 900 **17** 799 000

18 1 600 030 **19** 700 209

20 11 004 980 **21** 1 350 000 000

Rounding to whole numbers (pp. 7–9)

1

	Rounded to the nearest:	Highlight the digit to the right of the last required digit	Answer
518	ten	518	520
72 538	thousand	72 538	73 000
4 197	hundred	4 197	4200
634 116	ten thousand	634 116	630 000
2 457 101	million	2 457 101	2 000 000
67 452	hundred	67 452	67 500

2

	Round to the nearest ten	Round to the nearest hundred	Round to the nearest thousand
7 813	7 810	7 800	8 000
23 465	23 470	23 500	23 000
817 337	817 340	817 300	817 000
1 949	1 950	1 900	2 000
96 726	96 730	96 700	97 000
3 455 752	3 455 750	3 455 800	3 456 000
531 425	531 430	531 400	531 000

3 40 **4** 830

5 16 460 **6** 5 150

7 400 **8** 13 600

9 5 371 100 **10** 712 200

11 2 000 **12** 185 000

13 1 000 **14** 27 000

15 50 000 **16** 64 330 000

17 4 120 000 **18** 110 000

19 13 000 000 **20** 2 000 000

21 56 000 000 **22** 10 000 000

23

Original number	Rounded number	This was rounded to the nearest:
947	950	Ten
8 276	8 000	Thousand
12 846	12 850	Ten
367 482	36 500	Hundred
452 791	450 000	Ten thousand
24 613 094	24 600 000	Hundred thousand
490 578 935	500 000 000	Hundred million
1 642 542 895	2 000 000 000	Billion

ISBN: 9780170477710

Integers (pp. 10–11)

Adding and subtracting

1 $3 - 7 = -4$

2 $-2 - 3 + 11 = 6$

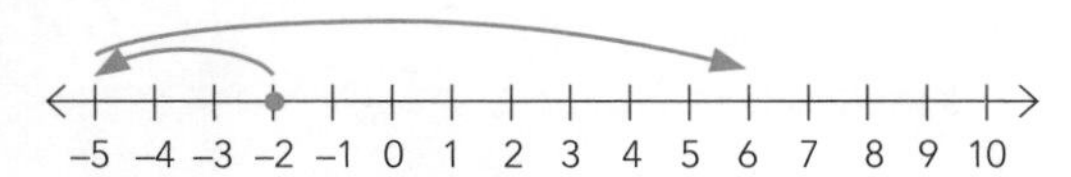

3 $-1 - 5 - (-2) = -4$

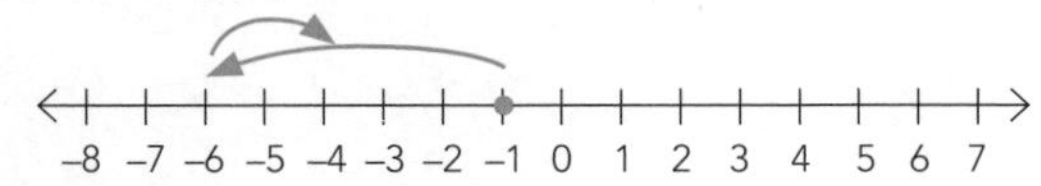

4 $-40 + 55 - 70 = -55$

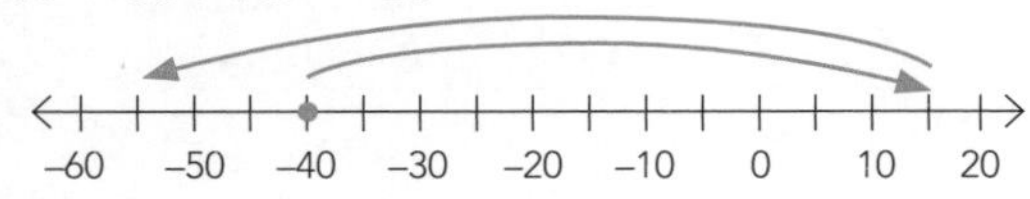

5 –36 **6** –9

7 –6 **8** –2

9 –90 **10** –214

11

Smallest / Largest

–19	–10	–9	–1	0	9	10	19

12

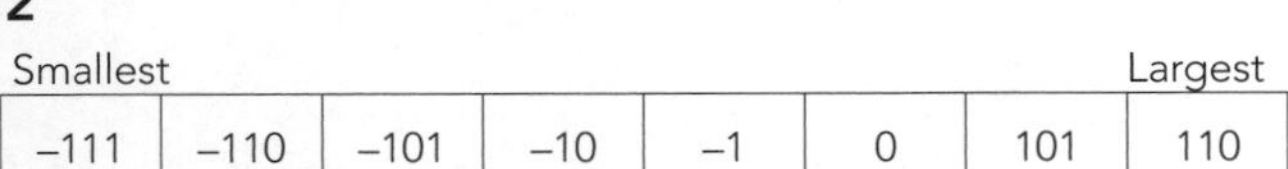

Smallest / Largest

–111	–110	–101	–10	–1	0	101	110

13

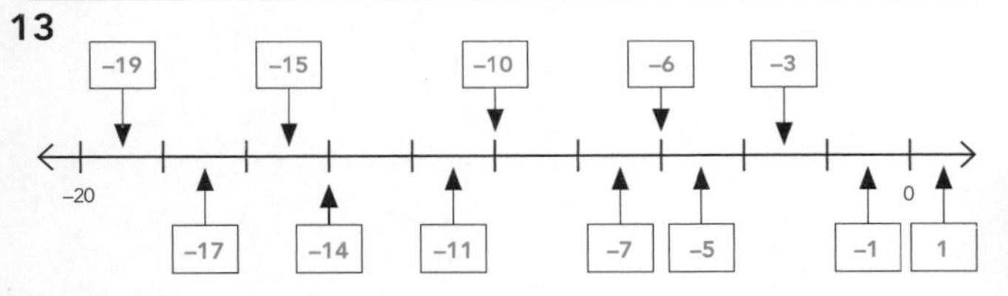

14 –6°C **15** $42

16 –15°C

Powers (p. 12)

1 $5 \times 5 \times 5 \times 5 = 5^4$ **2** $3 \times 3 = 3^2$

3 $8 \times 8 \times 8 = 8^3$ **4** $7 \times 7 \times 7 \times 7 = 7^4$

5 $2^4 = 2 \times 2 \times 2 \times 2 = 16$

6 $3^5 = 3 \times 3 \times 3 \times 3 \times 3 = 243$

7 $6^3 = 6 \times 6 \times 6 = 216$

8 $9^4 = 9 \times 9 \times 9 \times 9 = 6561$

9 $2^3 \times 2^2 = 32$ **10** $3^3 \times 10^2 = 2700$

11 $5^4 \times 2^3 = 5000$ **12** $4^2 \times 10^5 = 1\ 600\ 000$

Words to calculations (p. 13)

1 $7.50 **2** $17.00

3 $34.00 **4** $6.20

5 $8.50 **6** $37.80

7 $91.00 **8** $42.60

Decimals (pp. 14–17)

1 19.7 **2** 1 651.3

3 10.62 **4** 98.98

5 0.90 **6** 0.137

7 2.34, 2.43, 3.24, 3.42

8 0.06, 0.16, 0.60, 0.61

9 9.00, 9.01, 9.10, 9.11

10 0.010, 0.011,0.101, 0.110

11

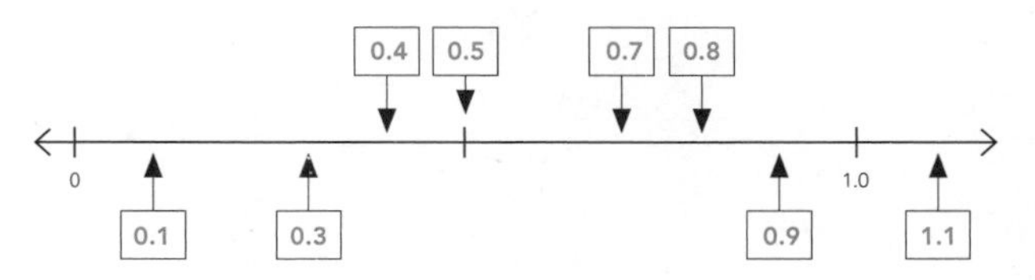

12

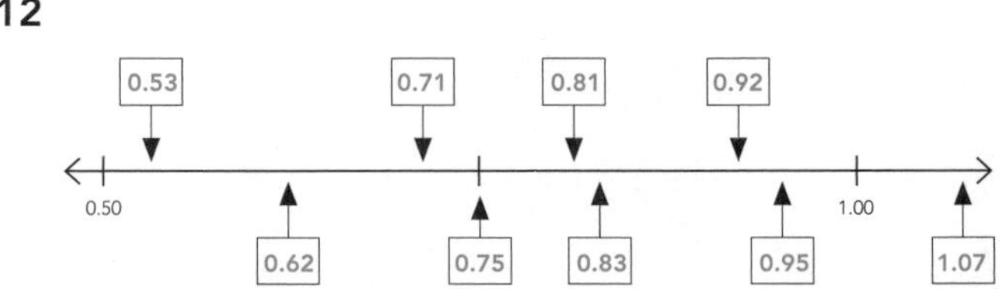

13

Original number	Rounded number	Rounded to how many dp?
4.814	4.81	2
95.571	95.6	1
251.19	251	0
1.0935	1.094	3
0.617	0.6	1
168.408	168.41	2
0.017	0.02	2

14

	Rounded to the nearest:	Highlight the digit to the right of the last required digit	Answer
29.372	1 dp	29.372	29.4
1.0375	2 dp	1.0375	1.04
132.6724	3 dp	132.6724	132.672
10.642	0 dp	10.642	11
9566.353	2 dp	9566.353	9566.35
0.3914	1 dp	0.3914	0.4
4.5271	3 dp	4.5271	4.527
0.0089	2 dp	0.0089	0.01

15

	0 dp	1 dp	2 dp
8.9421	9	8.9	8.94
16.738	17	16.7	16.74
9.3072	9	9.3	9.31
1.0997	1	1.1	1.10
0.7254	1	0.7	0.73
284.042	284	284.0	284.04

ISBN: 9780170477710

N

16 $0.56 \times 0.23 = 0.13$ (2 dp)
17 $1.2 \times 0.68 = 0.8$ (1 dp)
18 $10.05 \times 2.5 = 25.13$ (2 dp)
19 $0.88 \times 1.5 \times 0.49 = 0.647$ (3 dp)
20 $(3.61 + 1.8) \times 0.25 = 1.353$ (3 dp)
21 $18.4 \div 30 = 0.613$ (3 dp)
22 $\frac{4.165}{17} = 0.25$ (2 dp)
23 $\frac{244.5 + 6.35}{86.5} = 3$ (0 dp)

Fractions (pp. 18–20)

1 $\frac{1}{6}$
2 $\frac{3}{7}$
3 $\frac{1}{9}$
4 $\frac{5}{6}$
5 $1\frac{1}{3}$
6 $\frac{5}{5}$
7 $\frac{3}{5} = 0.6$, $\frac{5}{8} = 0.625$; $\frac{5}{8}$ is larger than $\frac{3}{5}$
8 $\frac{1}{3} = 0.\dot{3}$, $\frac{7}{20} = 0.35$; $\frac{7}{20}$ is larger than $\frac{1}{3}$
9 $1\frac{1}{5} = 1.2$, $1\frac{3}{25} = 1.12$; $1\frac{1}{5}$ is larger than $1\frac{3}{25}$
10 $4\frac{3}{4} = 4.75$, $4\frac{29}{40} = 4.725$; $4\frac{1}{3}$ is larger than $4\frac{13}{40}$
11 $\frac{13}{24}$
12 $1\frac{9}{10}$ or $\frac{19}{10}$
13 $2\frac{4}{21}$ or $\frac{46}{21}$
14 $4\frac{11}{20}$ or $\frac{91}{20}$
15 $1\frac{3}{8}$ or $\frac{11}{8}$
16 $5\frac{5}{8}$ or $\frac{45}{8}$
17 $\frac{4}{5}$
18 $21\frac{1}{2}$ or $\frac{43}{2}$
19 0.575
20 6.925
21 9.125
22 7.05
23 1.75
24 3.75
25 $\frac{7}{20}$
26 $\frac{7}{12}$
27 16
28 36
29 62.5 or $62\frac{1}{2}$
30 5.5 or $5\frac{1}{2}$
31 14
32 30
33 154
34 80
35 70
36 615
37 $28.50
38 16
39 304
40 5 194 000

Percentages (pp. 21–26)

1 Percentage green 64%
Percentage white 36%
2 Percentage green 12%
Percentage white 88%
3 Percentage green 34%
Percentage white 66%
4 Percentage green 40%
Percentage white 60%
5 42.5%
6 40%
7 80%
8 66.6%
9 87.5%
10 92%
11 93.75%
12 16%
13 95%
14 72.2%
15 7%
16

Decimal	Fraction	Percentage
0.5	$\frac{1}{2}$	50%
0.6	$\frac{2}{5}$	60%
0.25	$\frac{1}{4}$	25%
0.2	$\frac{1}{5}$	20%
0.4	$\frac{2}{5}$	40%
0.8	$\frac{4}{5}$	80%
0.75	$\frac{3}{4}$	75%
$0.\dot{3}$	$\frac{1}{3}$	$33.\dot{3}\%$
$0.\dot{6}$	$\frac{2}{3}$	$66.\dot{6}\%$

17 6
18 23
19 5
20 84
21 18.72
22 112
23 74 (not 73.9)
24 153.03 million km^2
25 37 950
26 75
27 64.8
28 210 g
29 $644
30 62.4 km
31 72.8 kg
32 $150.23
33 131.44
34 483
35 $27.00
36 $40 250
37 $6608
38 67.2
39 166.4
40 216 g
41 $493
42 74.8 km
43 49.6 kg
44 $195.75
45 110.2 cm
46 $523 200
47 a $29.97 b $599.25
48 $4760

ISBN: 9780170477710

Rates (pp. 27–28)

1 $22.50
2 $1.71
3 a NZ$645 b US$48
4 A costs $\frac{3.23}{170}$ = $0.019 = 1.9 c per g

B costs $\frac{2.25}{90}$ = $0.025 = 2.5 c per g

So tube A is better value.

Check with your teacher if your reasoning is different.

5 400 g pot costs $\frac{5.20}{400}$ = $0.013 per g

2.5 kg tub $\frac{29.95}{2500}$ = $0.01198 per g

So the 2.5 kg tub is better value.

Check with your teacher if your reasoning is different.

Ratios (pp. 29–30)

1 $20:$70
2 $25:$30
3 96 g:144 g
4 20 mL:140 mL
5 1200 m:900 m
6 $30.75:$61.50
7 9 kg
8 414
9 Kahu $78, Leah $52
10 192
11 6
12 20

Calculations with money (pp. 31–41)

1

Total price (electronic price)	Swedish rounding (cash)	Which is cheaper?
$47.38	$47.40	Electronic
$5.32	$5.30	Cash
$0.47	$0.50	Electronic
$234.54	$234.50	Cash
$1290.72	$1290.70	Cash
$23.69	$23.70	Electronic
$315.98	$316.00	Electronic

2

Value of loan	Rate of interest	Interest per year	Number of years	Total value of interest
$2000	2%	$40	6	$240
$500	4%	$20	3	$60
$7000	3.5%	$245	4	$980
$12 000	8.5%	$1020	7	$7140

3 $150.00
4 $450.00
5 $9000.00
6 $37 500.00
7 a $600.00 b $1200.00 c $13 200.00 d $550.00
8 a $80.00 b $320.00 c $2320.00
9 a $180.00 b $3360.00
10 C $2000 + 4 x $2000 x 0.03
11 $6.30
12 $1.65
13 $76.95
14 $281.25
15 $11.50
16 $9.78
17 $90.85
18 $416.30
19 $10.35
20 $43.70
21 $1050.00 $8050.00

Q22–39: You may get different answers. If so, check with your teacher.

22 3.49 + 9.73 ≈ **3 + 10**
= 13
23 1.19 + 5.52 + 3.41 ≈ 1 + 6 + 3
= 10
24 38.03 – 17.57 ≈ 38 – 18
= 20
25 349 + 953 ≈ 300 + 1000
= 1300
26 4.2 + 9.1 x 4 ≈ 4 + 9 x 4
= 40
27 6.2 ÷ 2.5 + 7.35 ≈ 6 ÷ 3 + 7
= 9
28 342 + 12 + 9 ≈ 340 + 10 + 10
= 360
29 5488 + 634 – 115 ≈ 5500 + 600 – 100
= 6000
30 42 500 ÷ 7.045 ≈ 42 000 ÷ 7
= 6000
31 37.4 x 19 + 269 ≈ 40 x 20 + 270
= 1070
32 $\frac{11976}{361} \approx \frac{12000}{400}$
= 30
33 1.8(23.45 – 8.112) ≈ 2(23 – 8)
= 30
34 1469 ÷ (685 – 355) ≈ 1500 ÷(700 – 400)
= 5
35 (213 + 36.6) ÷ 5.2 ≈ (200 + 40) ÷ 5
= 48
36 $90 – $10 – $4 – $2 = $74
37 $50 – 2 x $4 – 3 x $10 = $12
38 Total: $8 + $17 + $12 = $37
Change: $13
39 Total: $25 + $10 + $7 = $42
Change: $58
40 $34.50
41 $38.25
42 $40.88
43 $48.75
44 $50.40
45 $35.91
46 $91.93
47 $113.85
48

Income	Rate	Hours	Amount
Regular	21.00	40	$840.00
Time and a half	31.50	7	$220.50
		Total gross pay	$1060.50
	Deductions	PAYE (10.5%)	$111.35
		Net pay	$949.15

ISBN: 9780170477710

N

49

Income	Rate	Hours	Amount
Regular	27.50	80	$2200.00
Time and a half	41.25	13	$536.25
		Total gross pay	$2736.25
	Deductions	PAYE (10.5%)	$287.31
		Net pay	$2448.94

50

Income	Rate	Hours	Amount
Regular	38.00	80	$3040.00
Time and a half	57.00	5.5	$313.50
		Total gross pay	$3353.50
	Deductions	PAYE (10.5%)	$352.12
		Net pay	$3001.38

51 **a** a = $820.81 – $54.75
b b = $766.06 + $598.40

52 **a** a = $220.66 – $39.95
b b = $119.25 + $348.90

53

Date	Transaction	Credit	Debit	Balance
6 Dec 22	Pay	$675.00		$698.23
8 Dec 22	Bike repair		$45.00	$653.23
8 Dec 22	Rent		$350.00	$303.23
8 Dec 22	Supermarket		$105.62	$197.61
11 Dec 22	B Mart		$96.89	$100.72
15 Dec 22	Babysitting	80.00		$180.72
16 Dec 22	Dairy		$13.25	$167.47

54 D

55

Date	Transaction	Credit	Debit	Balance
3 Nov 22	B Bart		$56.80	$32.55
5 Nov 22	Auto Supplies		$125.66	–$93.11
8 Nov 22	Cafe		$8.50	–$111.61
8 Nov 22	Pay	$328.50		$216.89
9 Nov 22	Supermarket		$85.36	$131.53

Mixing it up (pp. 42–43)

1 **a** Four million, nine hundred and twenty thousand, six hundred and forty-eight.
b 4 921 000 **c** 984 130

2 **a** ± 4.7°C **b** 13.916 million km^2
c 1 900 000 million km^2
d 600 000 **e** 490 g
f 22 kg **g** $19 500.00

3 **a**

Income	Rate	Hours	Amount
Regular	23.00	40	$920.00
Time and a half	34.50	5	$172.50
		Total gross pay	$1092.50
	Deductions	PAYE (10.5%)	$114.71
		Net pay	$977.79

b

Date	Transaction	Credit	Debit	Balance
5 Nov 22	Shoe Mart		$125.00	$230.69
7 Nov 22	Takeaways		$32.50	$198.19
11 Nov 22	Dairy		$13.55	$184.64
11 Nov 22	Pay	$977.79		$1162.43
15 Nov 22	Rugby club		$75.00	$1087.43

c **i** $75.00
ii $3725.00
iii Any value between $4200 and $4300.

Mathematical relationships (pp. 44–59)

Reading axes (pp. 44–45)

1 Major 5
Minor 1

2 Major 2
Minor 0.5

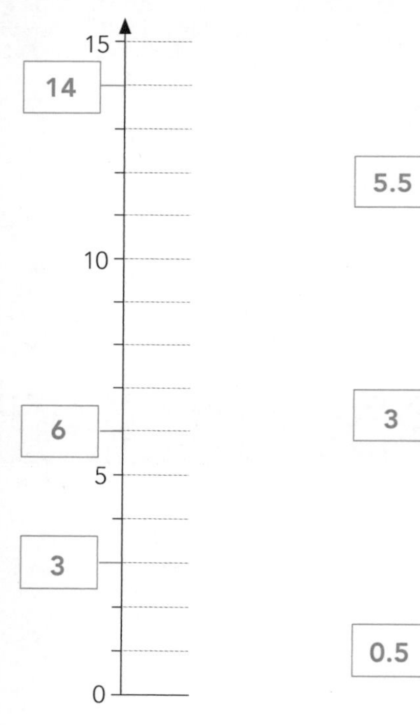

ISBN: 9780170477710

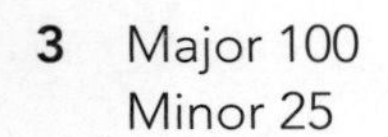

3 Major 100
Minor 25

4 Major 20
Minor 2

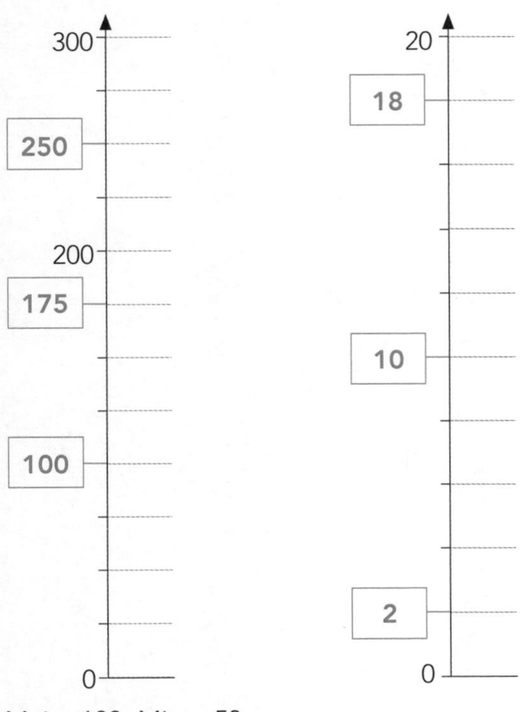

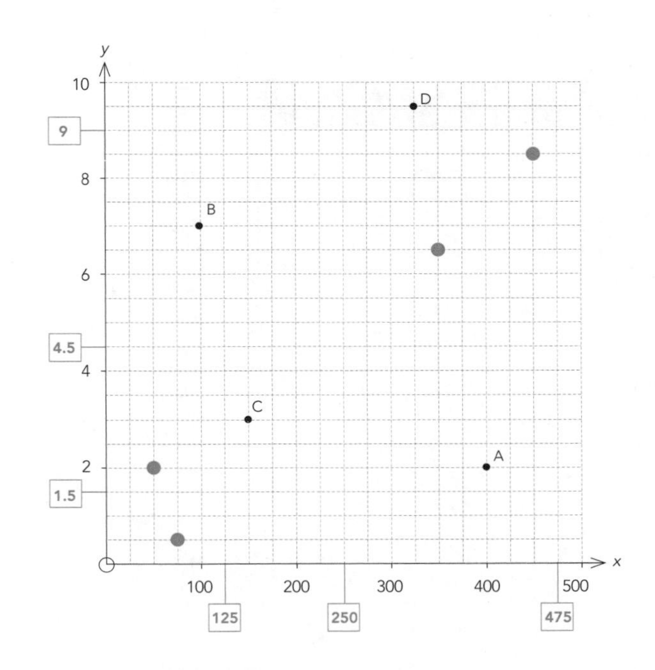

5 Major 100 Minor 50

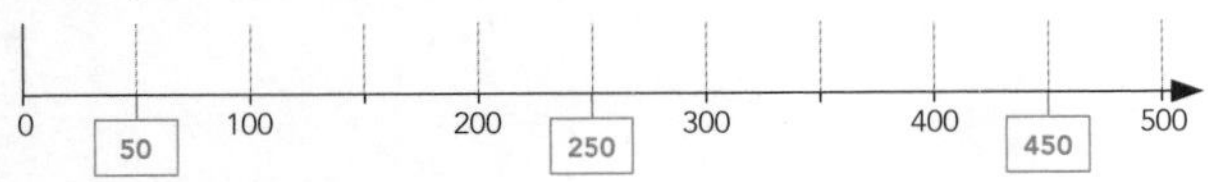

6 Major 50 Minor 10

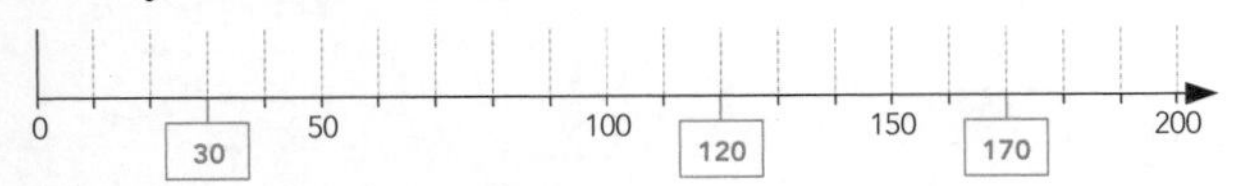

7 Major 1 Minor 0.25

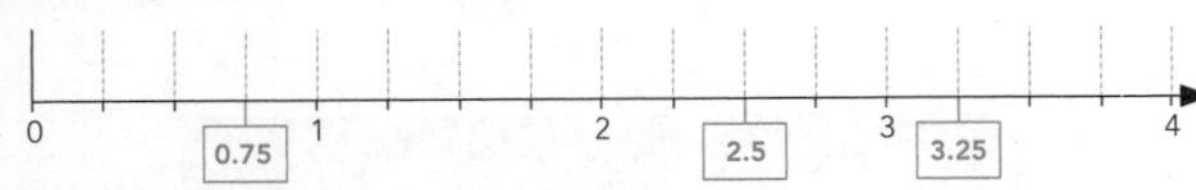

Coordinates (pp. 46–48)

1 a and **b** A(5, 15) B(10, 7) C(12, 23) D(2,6)

2 a and **b** A(400, 2) B(100, 7)
C(150, 3) D(325,9.5)

Linear relationships (pp. 49–52)

1

x	y
0	6
2	7
6	9
8	10

2

x	y
1	10
3	6
4	4
6	0

3

x	y
3	10
9	22
13	30
15	34

4

x	y
0	1.5
2	5.5
4	9.5
5	11.5

ISBN: 9780170477710

5

x	y
0	80
14	45
18	35
24	20

6

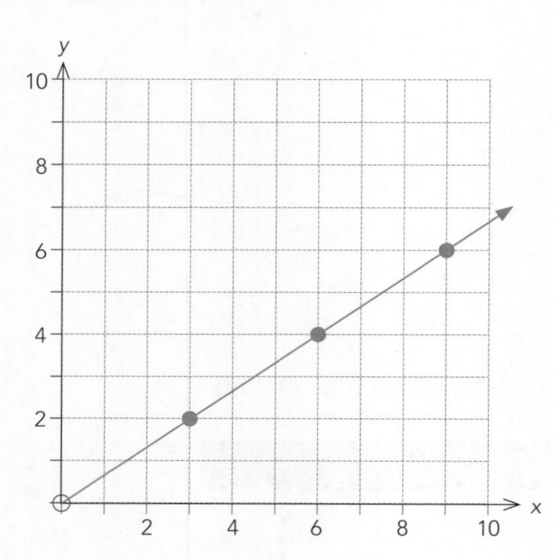

7

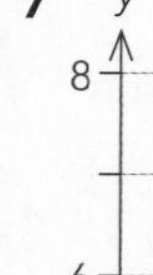

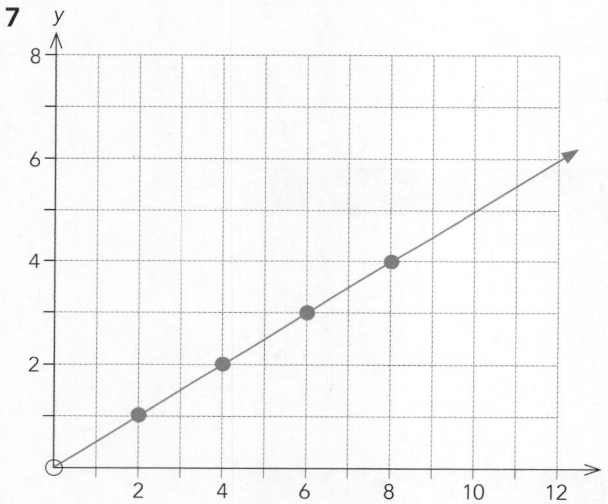

8

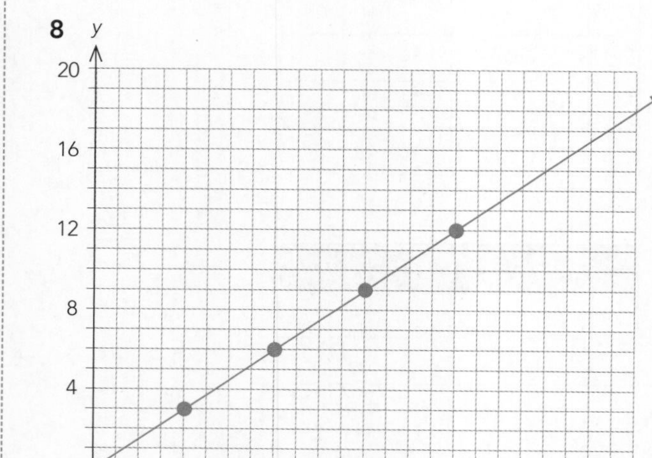

9

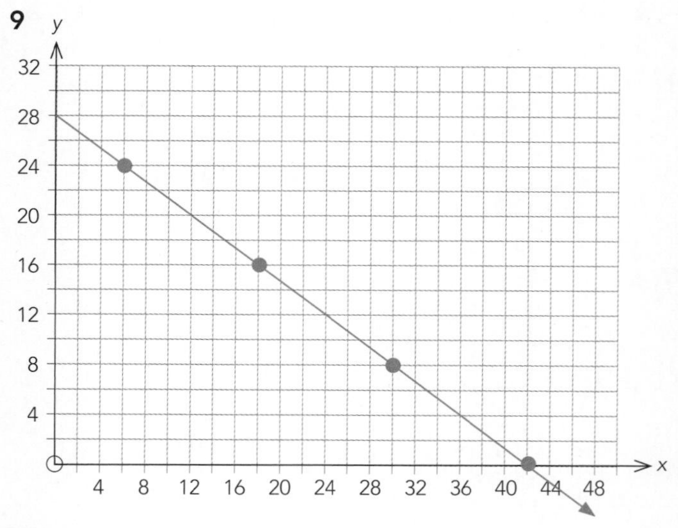

10

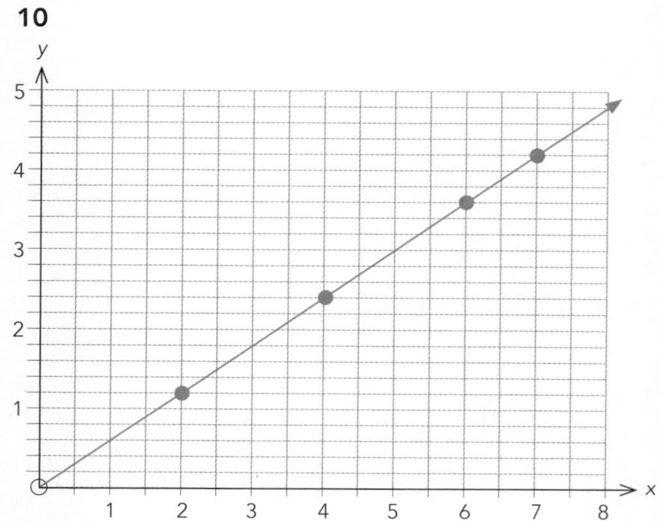

Applications (pp. 53–59)

1 **a**

Time (s)	Height (m)
0	0
10	16
20	32
30	48
40	64

b 8 m
c 72 m
d After 15 s
e After 25 s

2 **a** $5
b $8.00
c 4 kg
d 1.6 kg
e It means that 4.8 kg of kiwifruit will cost $12.

3 **a** 1 km
b 12 s
c As the time between lightning and thunder increases, the distance from the storm **increases/~~decreases~~**.
d When there is 6 seconds between the lightning and thunder, the storm is 2 km away.

ISBN: 9780170477710

4 **a** 2 min
b 44 cm
c 8 cm
d After 3.5 minutes there will be 50 cm of water in the tank.

5 **a** $560
b $320
c Fourth week
d As the weeks go by, the amount she owes her brother ~~increases~~/**decreases**.
e The y intercept would ~~increase/decrease~~/**stay the same**. The gradient would ~~stay the same~~/**become steeper**/~~become less steep~~.

6 **a** $60
b $330
c 4 hours
d As the number of hours spent on a job increases, his charge **increases**/~~decreases~~.
e The y intercept would **increase**/~~decrease/stay the same~~. The gradient would **stay the same**/~~become steeper/become less steep~~.

7 **a**

Mass (kg)	Cost ($)
1	2.50
2	5.00
3	7.50
4	10.00

b

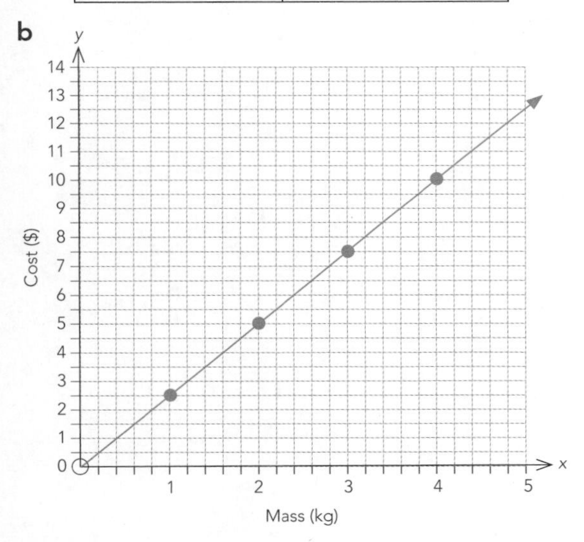

c $8.50 **d** 4.2 kg

8 **a**

Time (min)	Cost ($)
0	1.50
1	2.50
2	3.50
3	4.50
4	5.50

b

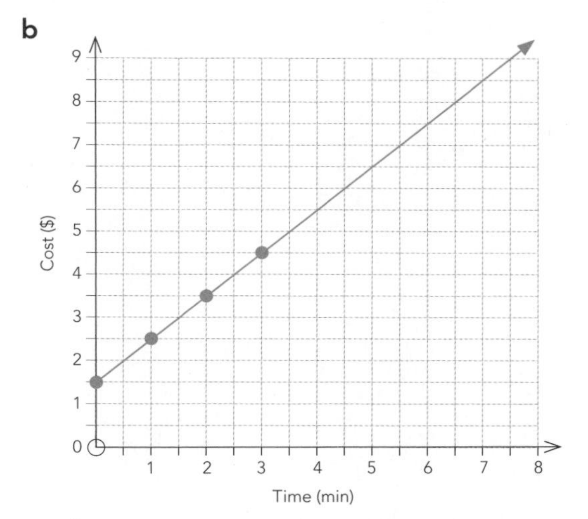

c $9.00 **d** 5.5 minutes

9 **a**

Time (min)	Distance from school (m)
0	2200
4	1900
8	1600
12	1300

b

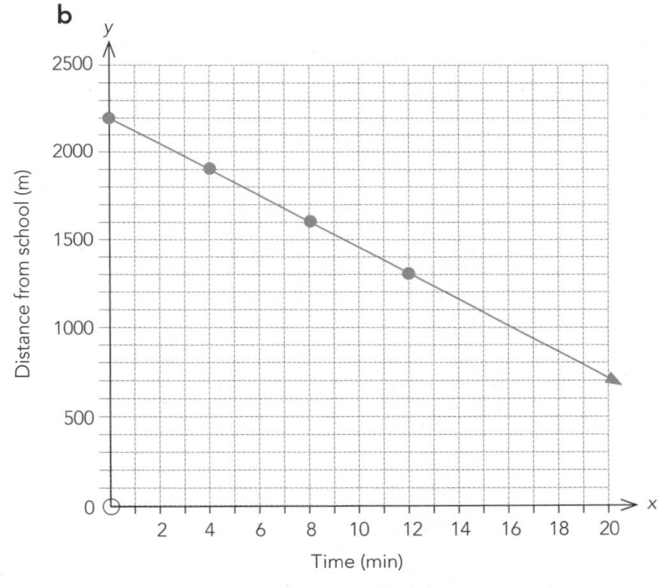

c 700 m **d** After about 18 minutes

Spatial properties and representations (pp. 60–78)

Transformation geometry (pp. 60–72)

1 Enlarged
2 Reflected
3 Rotated
4 Enlarged
5 Reflected
6 Translated
7 Rotated
8 Enlarged

ISBN: 9780170477710

9

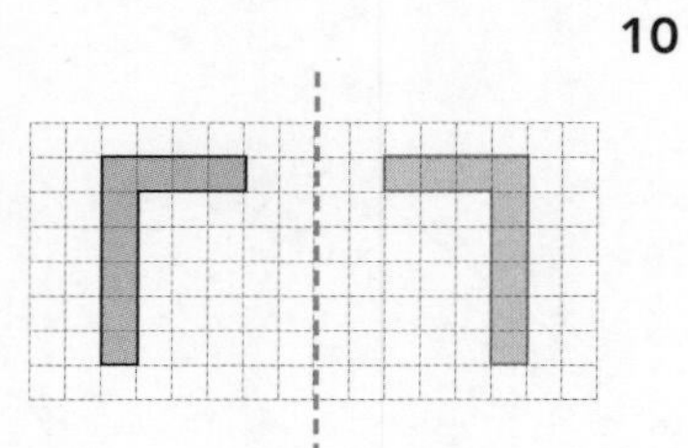

10

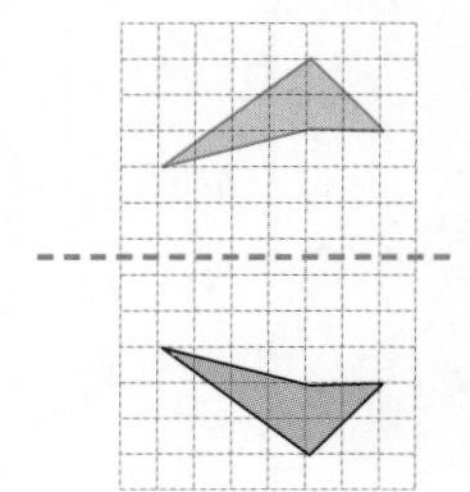

11

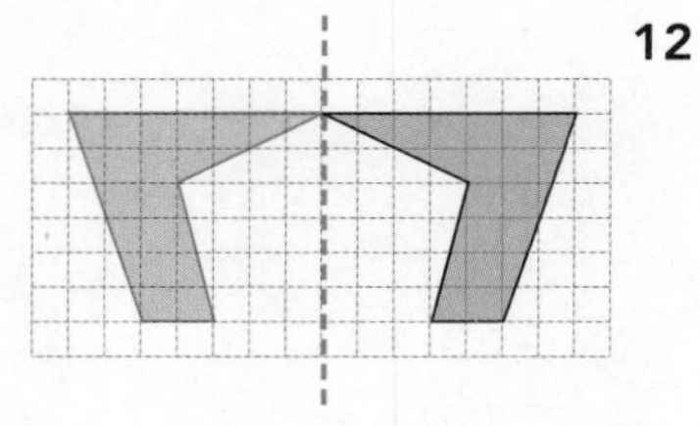

12

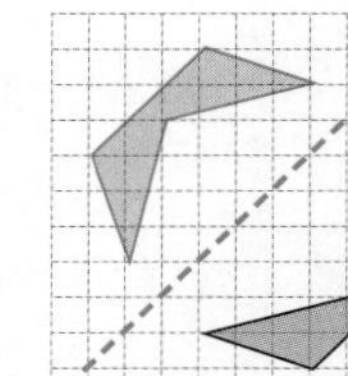

13 1 **14** 2
15 3 **16** 6
17 0 **18** 1
19 8 **20** 4
21 0 **22** 2
23 6 **24** 4
25 90° **26** 180°
27 270° **28** 270°
29 72° **30** 45°
31 36° **32** 30°
33 4 **34** 3
35 6 **36** 1
37 6 **38** 8
39 1 **40** 4
41 8 **42** 3

43

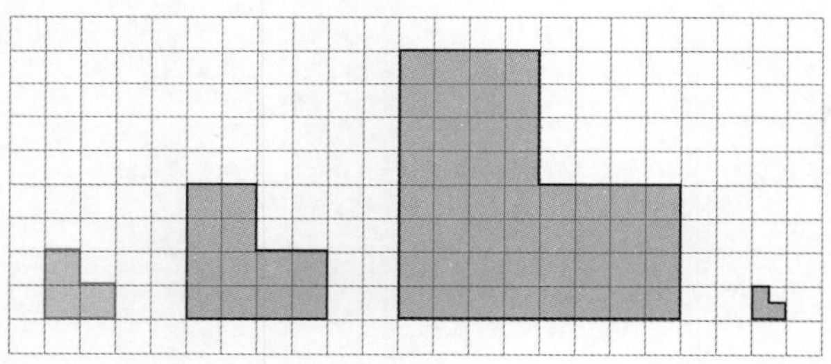

scale factor $= \frac{4}{2} = \frac{2}{1} = 2$ scale factor $= \frac{4}{1} = \frac{8}{2} = 4$ scale factor $= \frac{1}{2}$

44

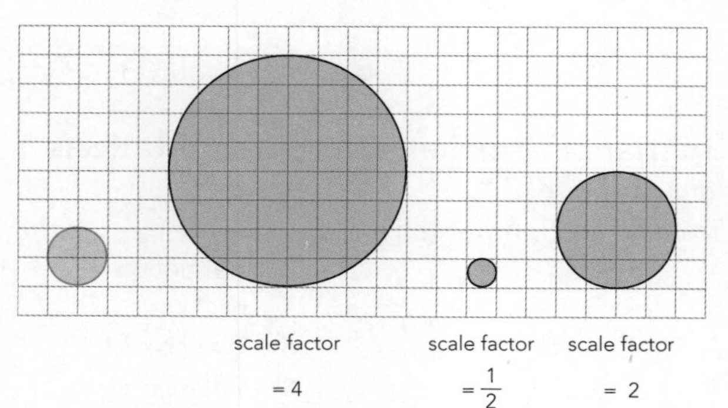

scale factor $= 4$ scale factor $= \frac{1}{2}$ scale factor $= 2$

45

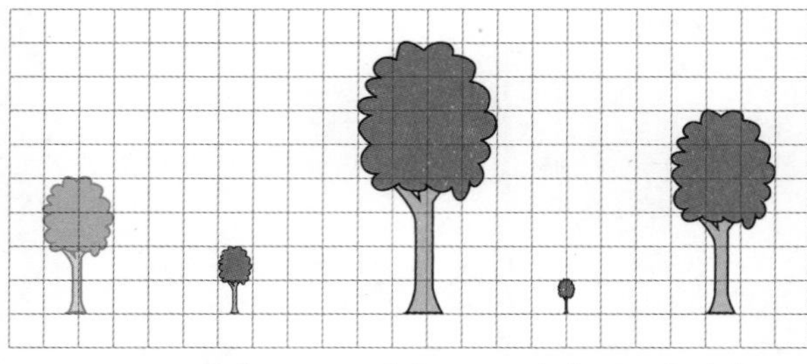

scale factor $= \frac{1}{2}$ scale factor $= 2$ scale factor $= \frac{1}{4}$ scale factor $= \frac{3}{2}$ or $1\frac{1}{2}$

46 a 80 cm² **b** 180 cm²
47 a 60 cm² **b** 375 cm²
48 The enlarged version will have 16 times the area of the original.
49 240 tiles

Different views of 3D shapes (pp. 74–75)

1 B
2 D
3 A
4 C
5 a 14 cubes **b** 15 faces
6 D
7

A	F	G	B	D	C	H	E

8 View B

Nets (pp. 76–78)

1

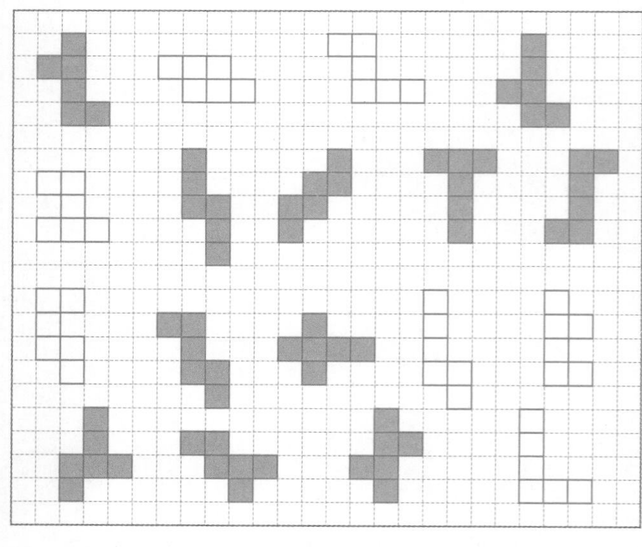

2 A

 ISBN: 9780170477710

3 a C

b

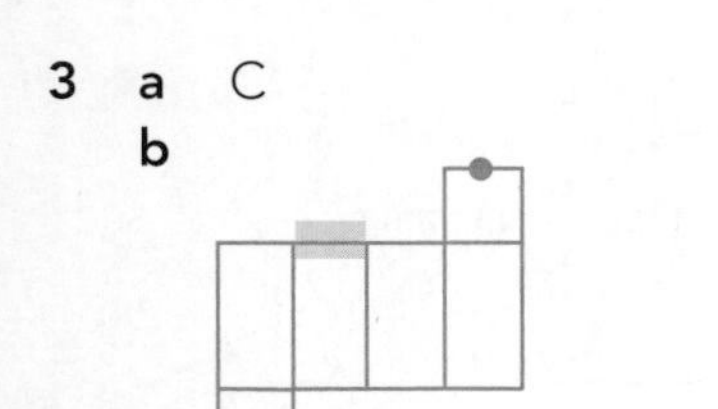

c

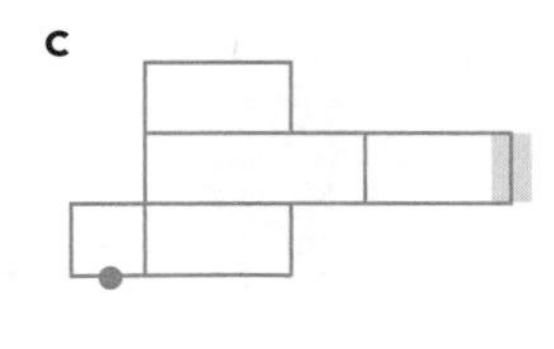

4 a

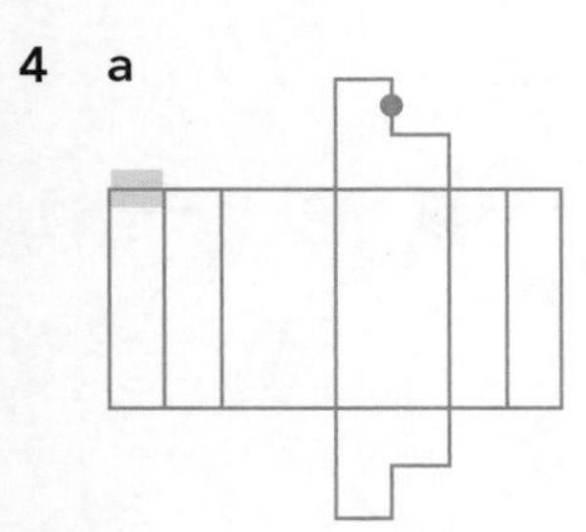

b

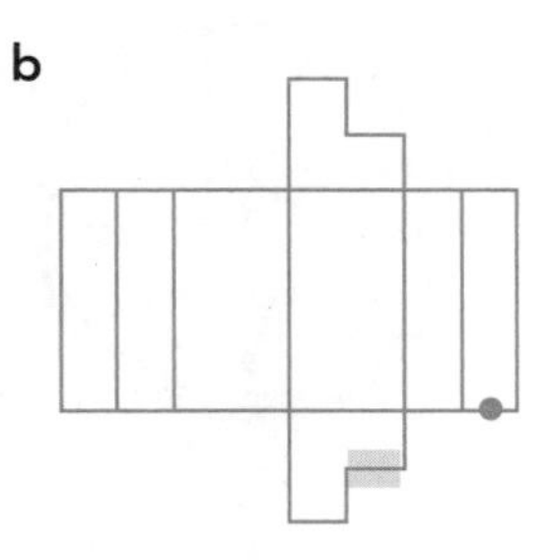

5

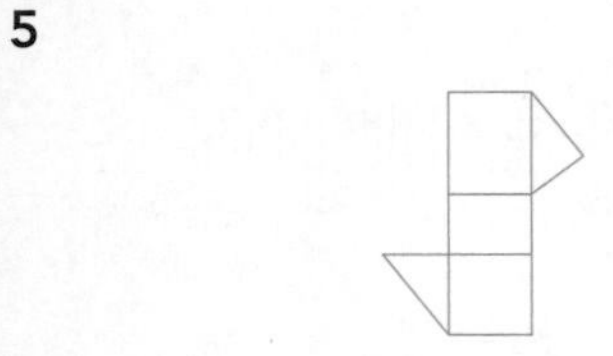

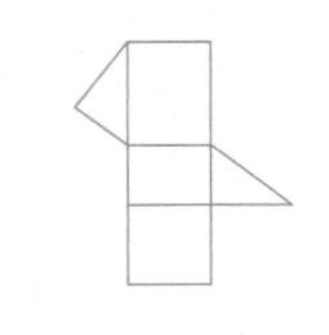

6

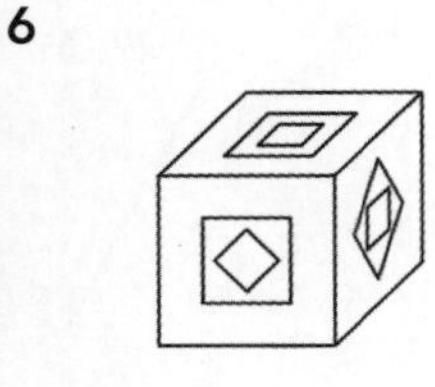

7

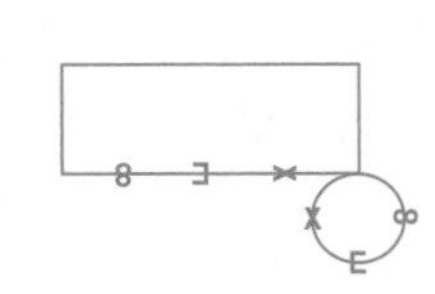

Location and navigation (pp. 79–89)

Directions (pp. 79–81)

1 a E b SE
c SW d N
e NW f S
g NE h W

2

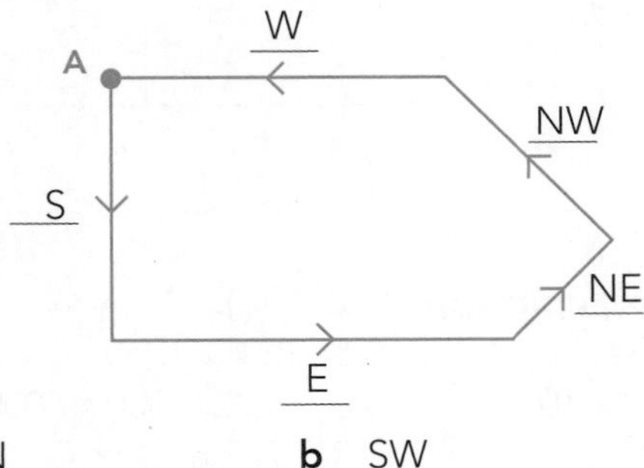

3 a N b SW

4 a NE
b Turn left out of Kimberley Rest Home and drive in a northeasterly direction to the town centre. Then turn left and drive in a northerly direction for four and half blocks.

Location: grid references (pp. 82–83)

1 Bus 2 Bike
3 D9 4 G8
5 Lion (or lioness, tiger, big cat etc)
6 Lion Encounter Meeting Point
7 Picnic table 8 Horse or zebra
9 I7 10 F9
11 I6 12 N10

Locating grid points (p. 84)

1 Bananas 2 Cherries
3 Apple 4 Strawberry
5 D7 6 I4
7 F2 8 E5

Scales (pp. 85–88)

1 a 10 km b 12 or 13 km
c 3 or 4 km d 11 or 12 km
e 25, 26 or 27 km f 18, 19 or 20 km
g 26, 27 or 28 km

2 a 4.1, 4.2 or 4.3 m b 6.0, 6.1 or 6.2 m
c 3.8, 3.9 or 4.0 m d 1.7, 1.8 or 1.9 m
e 1.5, 1.6 or 1.7 m f 1.9, 2.0 or 2.1 m
g 14.9, 15.0 or 15.1 m

3 a 11, 12, or 13 cm
b 3 or 3.5 cm c 3.5, 4 or 4.5 cm
d 3.5 or 4 cm e 3 or 3.5 cm
f 3 or 3.5 cm

Mixing it up (p. 89)

1 a Fiji b Manila
2 a A7 b B4
3 a W b N
c SE d NW
e SW f NE

4 a 1700, 1800 or 1900 km
b 600 or 700 km
c 2800, 2900 or 3000 km
d 2300, 2400 or 2500 km
e 3400, 3500 or 3600 km
f 10 300, 10 400, 10 500 or 10 600 km

Measurement (pp. 90–125)

Time (pp. 90–97)

1 a 480 ÷ 24 b 5 x 24 x 60
c 504 ÷ 24 ÷ 7

2 a 2 min b 1800 min
c 50 400 min d 2880 min

3 a 4 min b 48 h
c 6 h d 150 min
e 7200 s f 7.5 days
g 504 h h 4320 min

4 a 100 min b 1.5 days
c 190 s d 3 h
e 245 min f 460 s
g 3700 min h 172 801 s

5 a

95 min	0.1 day = 144 min	6000 s = 100 min	1.6 h = 96 min

Shortest Longest

95 min	1.6 h	6000 s	0.1 day

b

1.4 day = 2016 min	150 000 s = 2500 min	30 h = 1800 min	2000 min

Shortest Longest

30 h	2000 min	1.4 day	150 000 s

6 a 15:29 b 07:45
c 23:49 d 00:01
e 10:30 pm f 1:55 pm
g 4:09 am h 12:38 pm

7 a 21:08 b 03:04
c 23:59 d 10:45
e 12:00 f 00:00

8 a 3:11 pm b 6:50 pm
c 7:42 am d 10:27 pm
e 12:48 pm f 12:23 am

9 a 04:35 b 3:52 pm
c 6:21 pm d 2:02 pm
e 22:20 f 10:00 pm

10 a 07:00 b 13:10
c 23:40 d 12:45
e 19:10 f 16:45
g 2:00 pm h 10:35 am
i 10.35 pm

11 a 1 h 25 min b 3 h 41 min
c 1 h 55 min d 16 h 31 min
e 13 h 32 min f 3 h 39 min

12 a mid August b mid May
c mid August d mid September
e October 2022 f January 2024
g July 2023 h March 2023

13 a Every 15 minutes b 26 minutes
c 07:33

14 a Every 20 minutes b 26 minutes
c 11:03 am d 37 minutes
e Long Beach

15 a 49 minutes b 3:55 pm
c 16:32 or 4:32 pm d 16:25 or 4:25 pm

Scales (pp. 98–101)

1 A = 0.3 cm
B = 2.9 cm
C = 5.6 cm

2 D = 48 mm
E = 74 mm
F = 102 mm

3 48 mL
4 21°C
5 –33°C
6 36.6°
7 90 km/h
8 1.19 kg
9 130 amps
10 29 volts
11 1.115 m

12

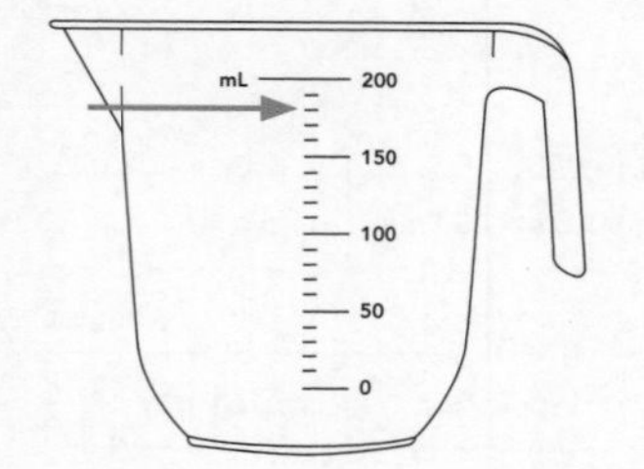

13

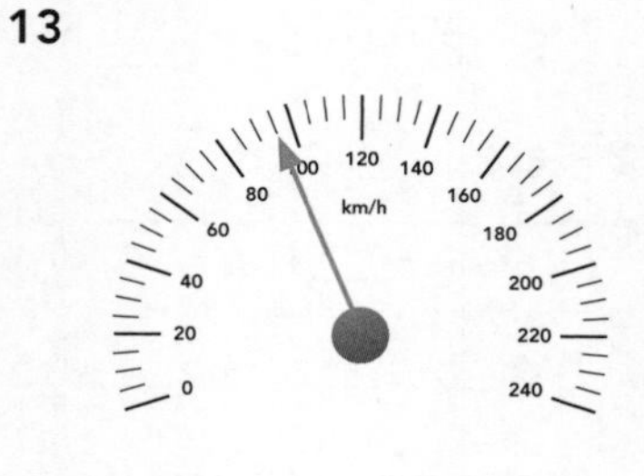

14

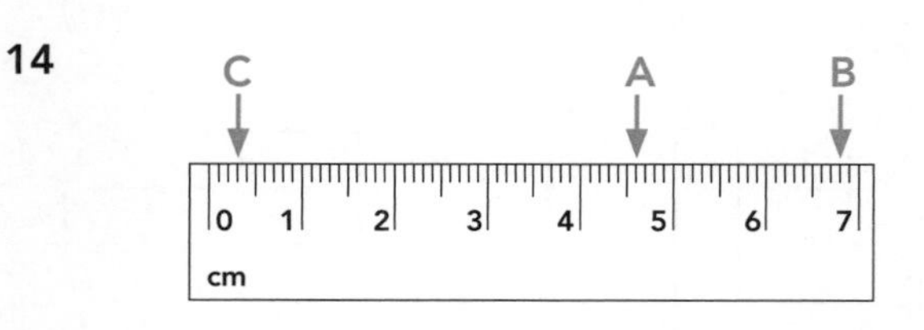

15 16

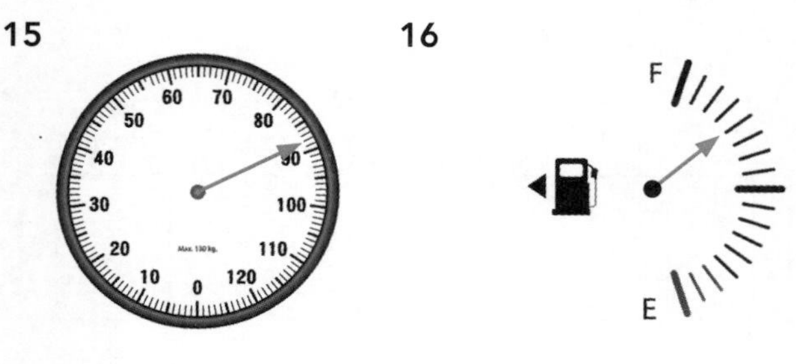

17

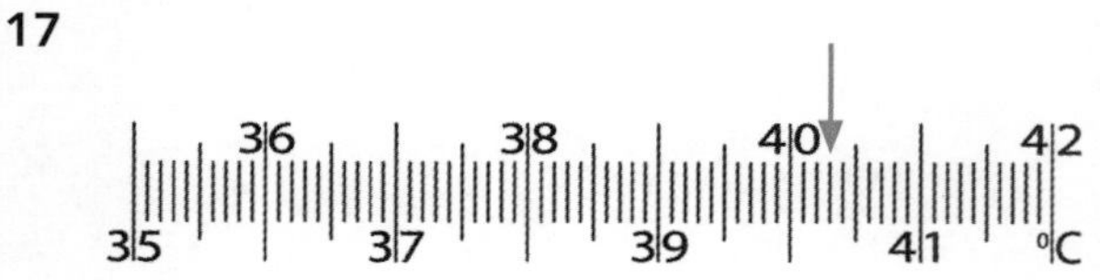

18 19

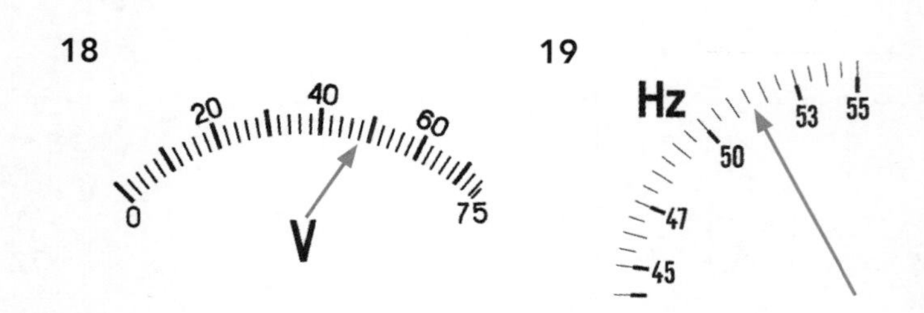

20

21

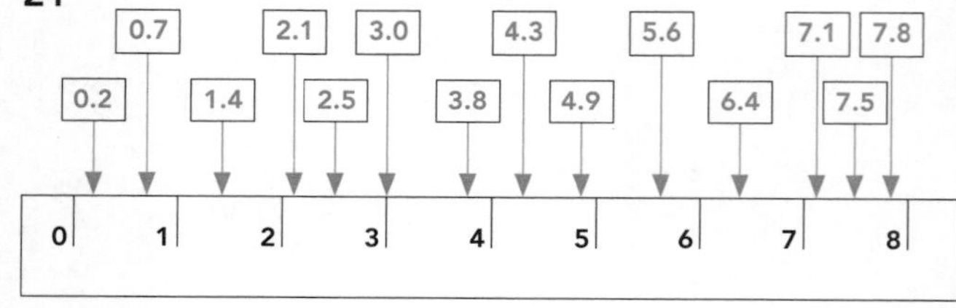

Converting units (pp. 102–106)

1 a 500 ÷ 100 b 2.7 x 1000
c 1.8 x 1000 d 0.6 x 1000 x 1000

2 600 cm
3 70 cm
4 2100 m
5 8.5m
6 5300 mm
7 30 000 cm
8 2 km
9 140 mm
10 3500 m
11 70 cm
12 68 m
13 6 cm
14 50 mm
15 0.8 km
16 3.5 m
17 0.9 cm
18 160 mm
19 4.6 m
20 18 000 m
21 23 cm
22 0.2 m
23 1.5 km
24 0.04 m
25 1.01 m
26 11 000 cm
27 3.6 km

 ISBN: 9780170477710

28 a 3.4 x 1000 b 27 000 ÷ 1000 ÷ 1000
c 0.8 x 1000 x 1000
29 2000 g 30 5 kg
31 3000 kg 32 45 g
33 15 kg 34 700 kg
35 38 000 g 36 8000 kg
37 76 g 38 7.9 kg
39 1500 kg 40 0.840 t
41 600 mg 42 0.39 kg
43 0.029 t 44 0.935 mg
45 60 g 46 89 000 g
47 30 g 48 0.7 t
49 30 kg 50 8 500 g
51 0.4 kg 52 2 million g
53 14.2 t 54 0.06 t
55 9 L 56 1500 mL
57 80 mL 58 24 L
59 9000 mL 60 7.3 L
61 42 L 62 5600 mL
63 29 000 mL 64 0.55 L
65 0.02 L 66 200 mL

Perimeter (pp. 107–110)

1 22 m 2 12 cm
3 160 mm 4 122 mm
5 24 km 6 10.5 km
7 144 mm 8 32.3 cm
9 12.5 m 10 13.7 km
11 250 cm or 2.5 m 12 410 mm
13 231 mm 14 30 cm
15 17.4 cm or 174 mm 16 269 mm
17 120 mm 18 246 cm or 2.46 m
19 201 cm or 2010 mm 20 9.6 cm

Area (pp. 111–116)

1 16 cm^2 2 24 km^2
3 28 mm^2 4 9 m^2
5 24 cm^2 6 25 m^2
7 45 km^2 8 17.64 mm^2
9 11.5 m^2 10 0.36 km^2
11 8.55 cm^2 12 0.0004 km^2
13 36 cm^2 14 56 m^2
15 15 km^2 16 10 km^2
17 36.9 cm^2 18 6750 mm^2
19 3.75 m^2 20 0.3024 km^2
21 245 92 cm^2 22 3756.5 m^2
23 14 m^2 24 13.5 km^2
25 1755 mm^2 26 20 m^2
27 60 cm^2 28 7.875 m^2
29 1012.5 mm^2 30 55 cm^2
31 2.21 m^2 32 29.11 cm^2

Volume (pp. 117–120)

1 12 cm^3 2 12 cm^3
3 27 cm^3 4 16 cm^3
5 48 cm^3 6 64 cm^3
7 120 cm^3 8 126 cm^3
9 36 cm^3 10 7.5 m^3
11 3648 cm^3 12 122 500 mm^3
13 22 m^3 14 614 125 mm^3
15 350.336 cm^3 16 0.6435 m^3

Using rates (pp. 121–123)

1 a 42 minutes b 48 minutes
2 140 beats per minute 3 12.5 hours
4 33 000 cm or 330 m or 3.3 km
5 $17.97 6 35 cookies
7 8 750 000 snaps 8 18 jars
9 8820 spiders 10 60.48 L
11 64 hours 12 51 km/h
13 14 minutes and 24 seconds
14 128 mm

Mixing it up (pp. 124–125)

1 a 4 hours 48 minutes
b 28.8 L
c 3:17 pm
2 a 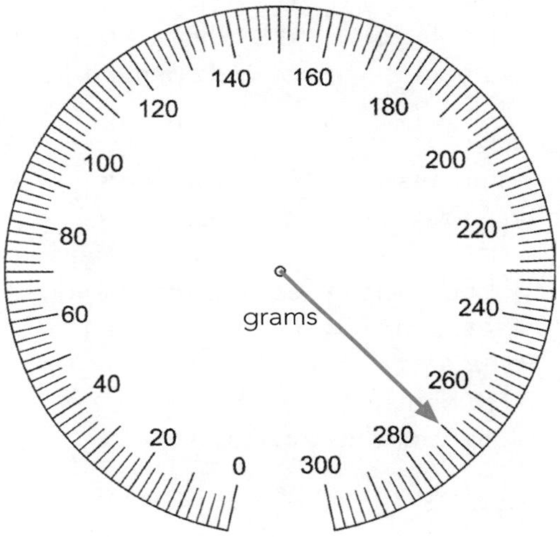
b 7 (not 7.4)batches
c 405 g d 2:50 pm
3 a 15.4 m b 4 rolls
c 4 duck eggs
4 a Every 30 minutes b 31 minutes
c 11.17
5 a 2.4 m^2 b 1.4 m
c 1.4 x 1.5 x 0.2 m^3

Statistics and data (pp. 126–146)

Types of variables (p. 126)

1 Descriptive 2 Continuous
3 Discrete 4 Descriptive
5 Continuous 6 Discrete
7 B, E 8 A, D
9 C, F

Data display (pp. 127–140)

1 a Red b Green
c Blue or white d $\frac{2}{3}$

ISBN: 9780170477710

2 a Sushi b Salad
c Filled rolls and pies
d 6 students

3 a 4 students b 7 or 8 students
c Technology d 40%

4 a 10 b Jam

5 a $2 b More than $2.

6 a 10 students b 20 students
c 15%

7 a Gardening b Cleaning cars
c $240

8 a 8 students b 21 students
c $\frac{3}{21}$ or $\frac{1}{7}$

9 a 14.3% b 5609 votes
c 2960 votes

10 a Toby b 214 babies
c Toby

11 a 14.6°C b July and August
c 6 months

12 a 69 athletes b 2016
c 184 athletes d 1988

13 a 13 students b 7 students
c $\frac{8}{44}$ or $\frac{2}{11}$

14 a 7 students b 26 students
c 80% to 100% d 40 students

15 a 7 students b Between 3 and 4 hours
c Nobody watched screens for a period of 6 to 7 hours.
d 9 students

16 a 30 students b Mars bar
c Moro, Crunchie and Snickers
d $\frac{12}{30}$ or $\frac{2}{5}$

17 a Year 11
b There are more dots towards the right for Year 11 than there are for Year 12.
or
There are more dots towards the left for Year 12 than there are for Year 11.
c 5 students

18

Graph	Name	Suitable for
	Pie graph	Descriptive/Discrete data
	Bar graph	Descriptive/Discrete data
	Pictograph	Descriptive/Discrete data
	Dot plot	Descriptive/Discrete data
	Line graph	Discrete/Continuous data
	Histogram	Continuous data
	Strip graph	Descriptive/Discrete data

Data analysis (pp. 141–144)

1 a $\text{Mean} = \frac{6+4+2+8+9+1+3+7+9+3}{10} = 5.2$

b $\text{Mean} = \frac{5+0+7+3+4+2+9+1}{8} = 3.875$

c $\text{Mean} = \frac{12+11+16+19+14+12}{6} = 14$

d $\text{Mean} = \frac{25+31+27+20+42+31+27}{7} = 29$

2 a Median = 4 b Median = 16.5
c Median = 11.5 d Median = 7
e Median = 4.5

3 a Mode = 2 b Modes = 4 and 7
c No mode d Modes = 1 and 4
e No mode f Mode = 4

4 a Range = 11
b Range = 21

5 a Smallest number = 5
b Largest number = 93

6 11C: Range = 10 – 3 = 7 11D: Range = 10 – 1 = 9
∴ 11D has a larger range (9) than 11C (7), so the test results for 11D are more variable.

Unusual features (pp. 145–146)

1

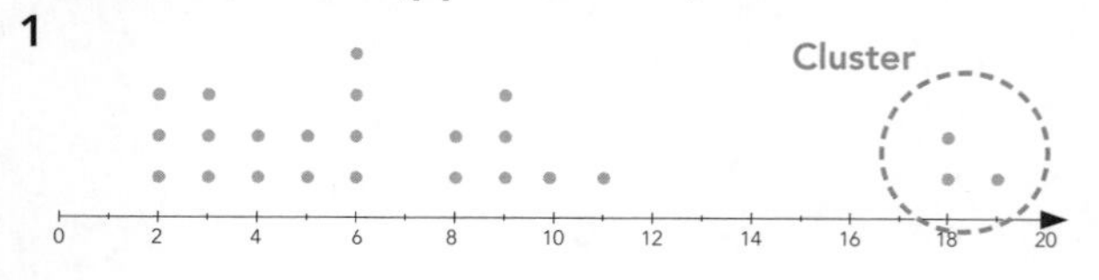

2

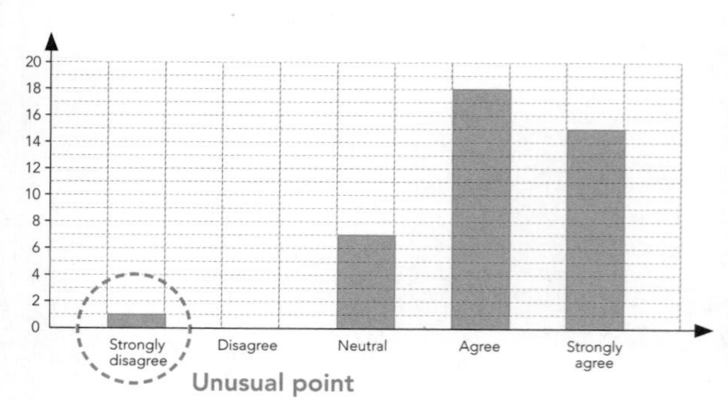

 ISBN: 9780170477710

3

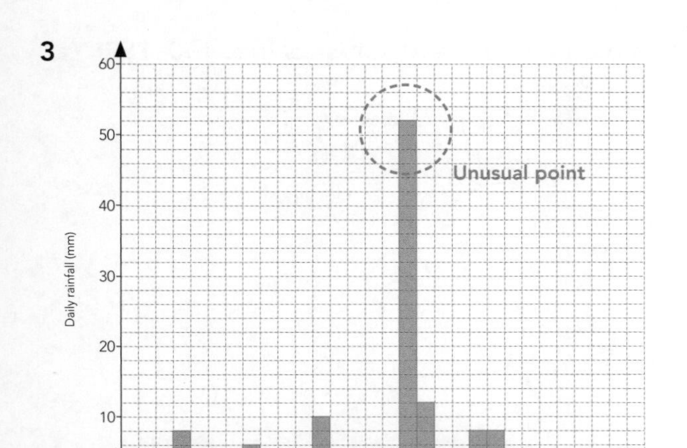

4 The 2s and 3s form a cluster.
5 9 is an unusual point.

Elements of chance (pp. 147–155)

The language of probability (p. 147)

1 **a** almost certain **b** guaranteed
c impossible **d** probable
e maybe **f** no way
g unlikely

2 **a** impossible **b** likely
c unlikely **d** certain
e even chance

The probability scale (p. 148)

Your answers may be different from some of these. If so, discuss them with your teacher.

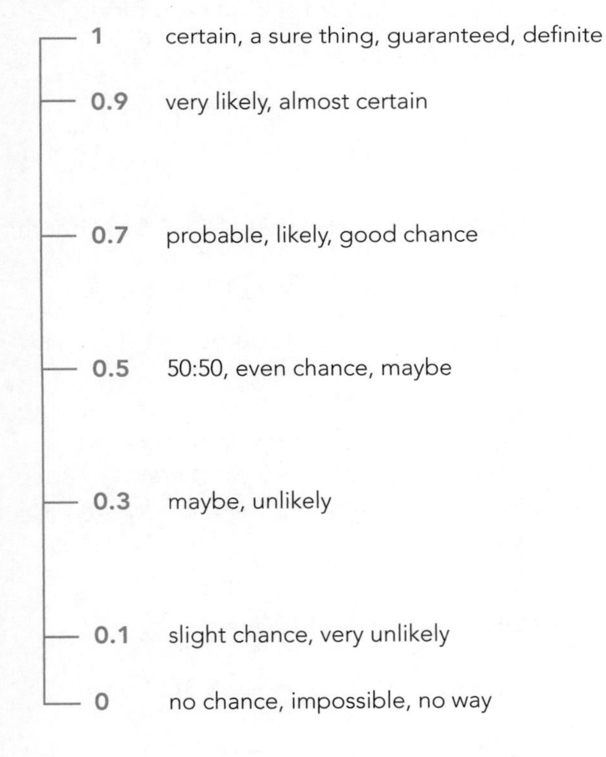

Calculating theoretical probability (pp. 150–151)

1 **a** $\frac{1}{6}$ or $0.1\dot{6}$ **b** $\frac{2}{6}$ or $\frac{1}{3}$ or $0.\dot{3}$
c $\frac{3}{6}$ or $\frac{1}{2}$ or 0.5 **d** 0
e $\frac{6}{6}$ or 1

2 **a** $\frac{3}{9}$ or $\frac{1}{3}$ or $0.\dot{3}$ **b** $\frac{2}{9}$ or $0.\dot{2}$
c $\frac{4}{9}$ or $0.\dot{4}$ **d** $\frac{5}{9}$ or $\frac{2}{3}$ or $0.\dot{5}$
e 0

3 **a** $\frac{2}{10}$ or $\frac{1}{5}$ or 0.2 **b** $\frac{7}{10}$ or 0.7
c $\frac{3}{10}$ or 0.3

Comparing probabilities (pp. 152–153)

1 $\frac{1}{5} = 0.2$ 19% = 0.19

More likely: 0.2

2 87% = 0.87 $\frac{11}{12} = 0.92$

More likely: $\frac{11}{12}$

3 53% = 0.53 $\frac{4}{7} = 0.57$

More likely: $\frac{4}{7}$

4 25% = 0.25 $\frac{2}{9} = 0.22$

More likely: 25%

5 $\frac{3}{7} = 0.43$ $\frac{6}{13} = 0.46$ 41% = 0.41

41% $\frac{3}{7}$ $\frac{6}{13}$

Least likely Most likely

6 Bag A: P(red) = 0.33
Bag B: P(red) = 0.43
Bag C: P(red) = 0.40
Most likely is bag B

7 Spinner A: P(green) = 0.38
Spinner B: P(green) = 0.33
Spinner C: P green) = 0.44
Most likely is spinner C

8 P(4 at top of die) = $\frac{1}{4}$ = 0.25

P(two heads) = $\frac{1}{4}$ = 0.25

No. The chances are the same.

9 $\frac{1}{216}$ = 0.0046, so tossing seven coins at once and getting all heads (probability 0.0078) is more likely.

10 $\frac{1}{383\,838}$ = 0.0000026, so tossing 18 coins at once and getting all heads (probability 0.0000038) is more likely.

Complementary events (p. 154)

1 0.6 **2** $\frac{5}{6}$ or $0.8\dot{3}$

3 $\frac{2}{3}$ or $\frac{6}{9}$ or $0.\dot{6}$ **4** $\frac{13}{25}$ or 0.52

5 $\frac{119}{637}$ or 0.19 (2 dp) **6** $\frac{6}{24}$ or 0.25

Risk (p. 154)

1 $\frac{13}{37}$ or 0.35 (2 dp)

2 **a** $\frac{1}{6}$ or $0.1\dot{6}$ **b** $\frac{2}{5}$ or 0.4

3 **a** $\frac{9}{54}$ or $0.1\dot{6}$ **b** $\frac{17}{102}$ or $0.1\dot{6}$

4 $\frac{12}{28}$ or 0.43 (2 dp)

Process ideas (pp. 156–165)

1 Formulate approaches to solving problems (pp. 156–157)

1 **a** Yes
5% interest on \$3500 = \$3500 x 0.05
= \$175 interest each year
So her total amount to be repaid = \$3500 + 2(175) = \$3850

b $\frac{\$3850}{104 \text{ weeks}}$ = \$37 (0 dp) or \$37.02

2 Yolk in a kiwi egg = 372 x 0.6
= 223.2 g
Yolk in a hen egg = 60 x 0.35
= 21 g
Number of hen yolks in a kiwi yolk = 223.2 ÷ 21 = 10.63 (2 dp), so he is correct because 10.63 rounds up to 11.

3 Deal 1: 4 plants cost 2(\$6.90 + \$3.45) = \$20.70
Deal 2: (4 x \$6.90) x 80% = \$22.08
So Deal 1 is \$1.38 cheaper.

4 Time taken = $\frac{3}{4} \times \frac{10}{3.6} = 2.08\dot{3}$ hours
= 2 hours + 0.083 x 60 min
= 2hours and 5 minutes

5 **a** Royal Raspberries pays \$50 per day plus \$1.50 for each punnet picked.

b Grower: Ruby Raspberries
Difference in pay: About \$15 (from the graph)

2 Use mathematics and statistics (pp. 158–159)

1 **a** 19

b Area = $\frac{1}{2}$ x 16 x 24 = 192 cm²

c Number of flags = $\frac{1530}{18}$ = 85

d \$17.34

e \$9.62

2 **a** 5.6 cm

b Histogram

c 29 shells

d

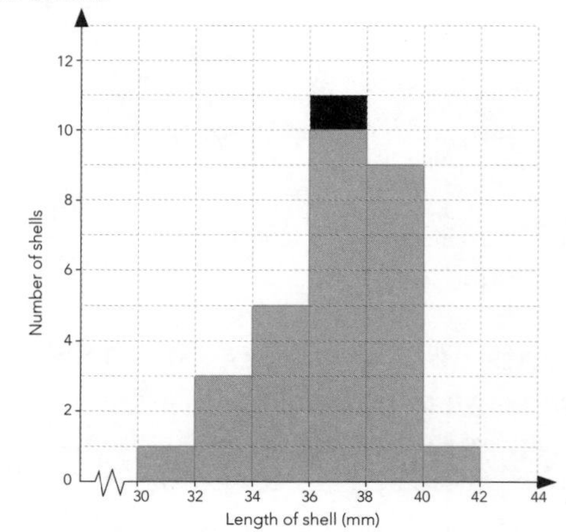

3 16 x \$0.8 + 15 x \$0.45 + \$1.75 = \$21.30

4 15 faces

3 Explain the reasonableness of responses (pp. 160–165)

1 Agree
Average profit per jar = 0.56 x \$1 + 0. 44 x \$2 = \$1.44
Proportion of profit from chutney and pickle
= $\frac{0.44 \times 2}{1.44}$ = 61% (0 dp)
So her friend is correct.

2 A Frankie's Firewood costs $\frac{\$400}{3.6}$ = \$111.11 per m³.
B Wallace's Wood costs \$95 + $\frac{\$36}{7.2}$ = \$100 per m³.
C Fergie's Fuel costs $\frac{\$108}{1.2}$ = \$90 per m³.
So option C is the lowest price per m³.

3 Disagree
The *y*-axis starts at 400, not 0. There were about 700 non-fiction books borrowed and just over 1000 fiction books borrowed. So there were not twice as many fiction books than non-fiction books borrowed.

4 Disagree
Probability of a win for A = $\frac{2}{5}$ = 0.4.
Probability of a win for B = $\frac{3}{8}$ = 0.375.
Probability of a win for C = $\frac{5}{12} = 0.41\dot{6}$.
So spinner C has the best chance of winning, and spinner B the worst.

ISBN: 9780170477710

5 No

There were 88 + 40 + 31 + 10 + 5 = 174 apples in the sample that were big enough for export. This represents $\frac{174}{380}$ = 45.8% of the sample. This is less than his target of 50%, so the sample suggests that he will not meet his target.

6 No

Subtracting 32 from 100 gives 68.

$\frac{68}{1.8} = 37.\dot{7}$

This rounds up to 38°C, so 100°F is closer to 38°C rather than 37°C.

7 Disagree

What has happened in the past will not affect his probability of throwing doubles in the future. The probabilty will remain at = $\frac{1}{6}$.

8 Yes.

His vision will extend to 45° above the horizontal (*x*-axis). That is the line that will pass through (1, 1), (2, 2), (4, 4), etc. The point (6, 4) wil lie below this line, so the ruru will be able to turn his head far enough to see the weta.

9 **a** The percentage of orders for at least 4 cups $= \frac{16 + 3 + 6 + 1 + 2}{92}$ = 30.4%. This is greater than 25% so it does support buying cup holders.

b The point where the line crosses the *y*-axis, (0, –400), means that if she doesn't sell any cups of coffee during a week, it will cost her $400. The point where the line crosses the *x*-axis, (200, 0), means that she needs to sell 200 cups of coffee to break even.

Numeracy practice (pp. 166–175)

Set 1 (pp. 166–167)

1 **a** 600 g

b 400 g

Disagree

The total percentage of the diet that is frogs, fish and insects is 30%, which is less than $33.\dot{3}$%, but not much less. The proportions shown are averages, so at times frogs, fish and insects might make up a third of the diet.

2 **a** Yes

Either: Old car used 14 L for 100 km.

New car uses 1 x $\frac{100}{12.8}$ = 7.8 L (1 dp) for 100 km, so the new car is more efficient.

or: New car can drive 12.8 km per litre of fuel.

Old car could drive only $\frac{100}{14}$ = 7.1 km (1 dp) using a litre of fuel, so the new car is more efficient.

b 120 mm

3 **a** ☑ The most common way of getting to school is by walking.

☑ Students are twice as likely to bike as they are to scooter to school.

☑ Most students get to school using something with wheels.

b

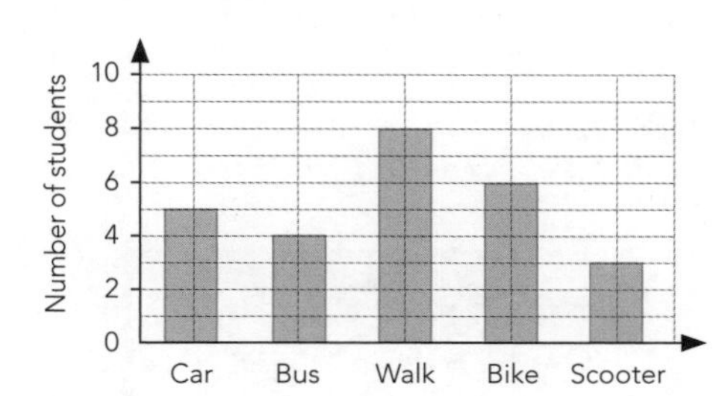

c 75–80%

d 4 (Remember, she has to travel to and from school.)

e 3

Set 2 (pp. 168–169)

1 **a**

b Celly, $674.10

2 **a** $600 **b** 225 Thai baht

c 1:50 pm

3 **a** NE **b** $456.50

c 12:32 pm **d** 22 km/h

e Disagree

Thirty times of arrival are on the graph. Of these, the ferry arrived early or on time on 16 occasions, but it was late on 14 occasions. This means it was late on only 47% (0 dp) of occasions, which is less than half.

Set 3 (pp. 170–171)

1 **a**

b 1.5 or 1.6 kg

c The gradient of the line would become steeper.

d P(A or N) = $\frac{4}{9} = 0.\dot{4}$

N

2

3 a $6.84\ m^2$

b

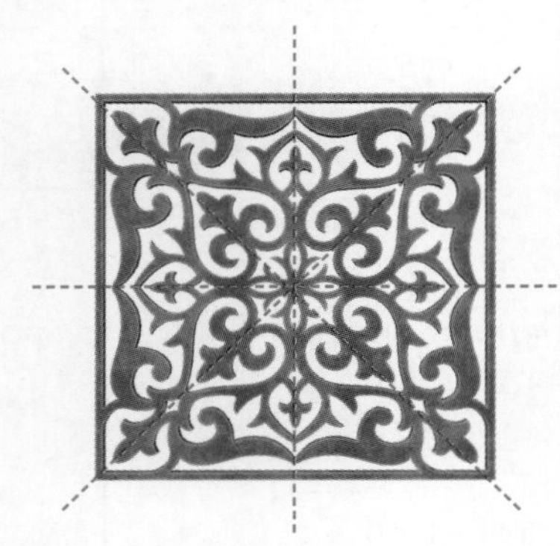

c 32 tiles

4 a NZ$1 = US$0.63

so $NZ 579 = $US 579 x 0.63

= $US 364.77

b I would convert 1:34 pm into 24-hour time: 13:34.

I would subtract 11:49 from 13.34 to give 1 hour 45 minutes.

I would convert 1 hour 45 minutes to decimal hours: 1.75 hours.

I would then divide the distance by the time:

$\frac{1060}{1.75} = 605.7$ km/h (1 dp)

Set 4 (pp. 172–173)

1 a

b $450.00

2 a 2 x 49.95 x 0.8 ÷ 12

b 170 cm

c 1151 cm

3 a 8:20 am

b 23 min

c The section from Rimu Lane to Totara Way takes 10 minutes, rather than 9 minutes on the first three buses.

The section from Daisy Lane to Elizabeth Street takes 6 minutes, rather than 5 minutes on the first three buses.

This is probably because the traffic gets busier from 8 am onwards because many people travel to work then. Also, with more people travelling, getting on and off the bus will take longer.

4 a $5.50 b 55 km/h

Set 5 (pp. 174–175)

1 a 25.24% (2 dp)

b 164 cm

c 39.50 cm

d Either:

Agree, because 87 women had femur lengths between 45 and 55 cm, and that is 58% of the women in the sample.

Or:

Can't tell for sure, because the sample was taken from New Zealand women who might be taller or shorter on average than 'most women' from round the world.

e

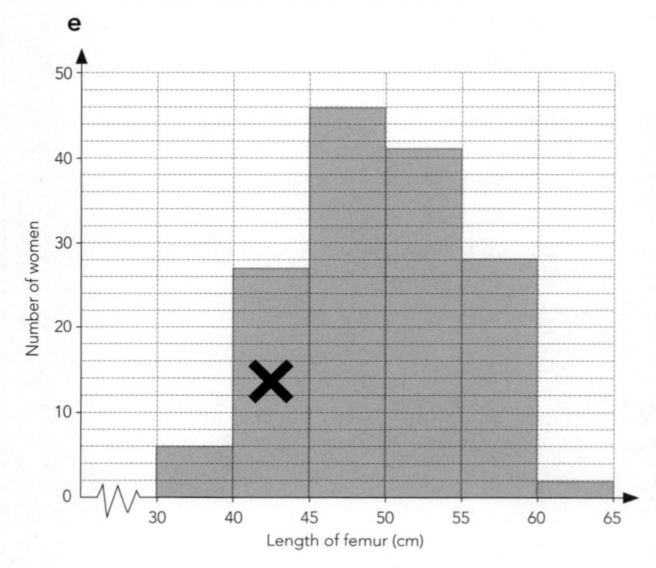

2 a

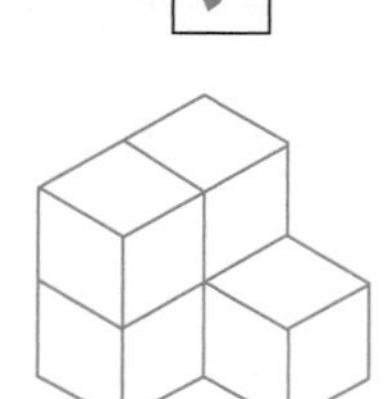

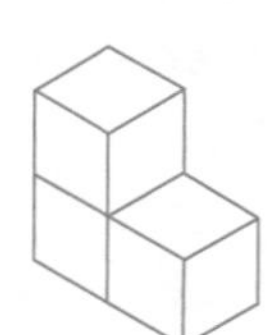

b 18 faces

3 a The normal entry cost for 2 adults ($48) and two children ($26) is $74. The family pass is $60, so you only save $14, not $15 as advertised.

b 3798 tourists

c 7.4 cm, 7.5 cm or 7.6 cm

 ISBN: 9780170477710